Statistics 2

Edexcel AS and A Level Modular Mathematics

Greg Attwood
Alan Clegg
Gill Dyer
Jane Dyer

Contents

About this book

This book is designed to provide you with the best preparation possible for your Edexcel S2 unit examination:

- This is Edexcel's own course for the GCE specification.
- Written by senior examiners.
- The LiveText CD-ROM in the back of the book contains even more resources to support you through the unit.

Finding your way around the book

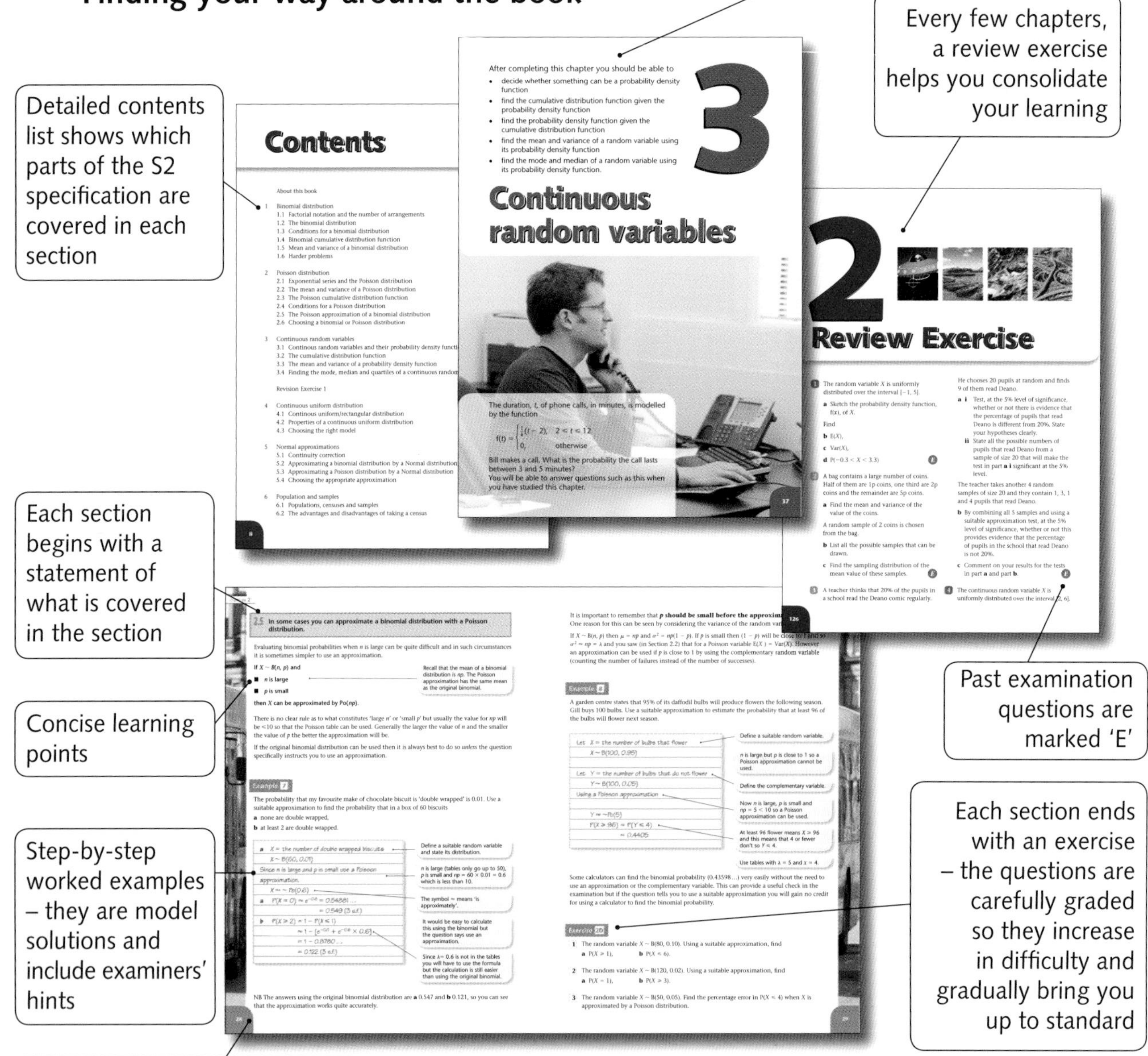

Each chapter has a different colour scheme, to help you find the right chapter quickly

Each chapter ends with a mixed exercise and a summary of key points.

At the end of the book there is an examination-style paper.

LiveText software

The LiveText software gives you additional resources: Solutionbank and Exam café. Simply turn the pages of the electronic book to the page you need, and explore!

LiveText

2.5 In some cases you can approximate a binomial distribution with a Poisson distribution.

Evaluating binomial probabilities when n is large can be quite difficult and in such circumstances it is sometimes simpler to use an approximation.

If $X \sim B(n, p)$ and

- n is large
- p is small

then X can be approximated by $Po(np)$.

Recall that the mean of a binomial distribution is np. The Poisson approximation has the same mean as the original binomial.

There is no clear rule as to what constitutes 'large n' or 'small p' but usually the value for np will be $\leqslant 10$ so that the Poisson table can be used. Generally the larger the value of n and the smaller the value of p the better the approximation will be.

If the original binomial distribution can be used then it is always best to do so *unless* the question specifically instructs you to use an approximation.

Example 7

The probability that my favourite make of chocolate biscuit is 'double wrapped' is 0.01. Use a suitable approximation to find the probability that in a box of 60 biscuits

a none are double wrapped,

b at least 2 are double wrapped.

a X = the number of double wrapped biscuits
$X \sim B(60, 0.01)$
Since n is large and p is small use a Poisson approximation.
$X \approx \sim Po(0.6)$

a $P(X = 0) = e^{-0.6} = 0.54881...$
$= 0.549$ (3 s.f.)

b $P(X \geqslant 2) = 1 - P(X \leqslant 1)$
$= 1 - [e^{-0.6} + e^{-0.6} \times 0.6]$
$= 1 - 0.8780...$
$= 0.122$ (3 s.f.)

Define a suitable random variable and state its distribution.

n is large (tables only go up to 50), p is small and $np = 60 \times 0.01 = 0.6$ which is less than 10.

The symbol $\approx$ means 'is approximately'.

It would be easy to calculate this using the binomial but the question says use an approximation.

Since $\lambda = 0.6$ is not in the tables you will have to use the formula but the calculation is still easier than using the original binomial.

NB The answers using the original binomial distribution are **a** 0.547 and **b** 0.121, so you can see that the approximation works quite accurately.

It is important to remember that **p should be small before the approximation is used**. One reason for this can be seen by considering the variance of the random variable.

If $X \sim B(n, p)$ then $\mu = np$ and $\sigma^2 = np(1 - p)$. If p is small then $(1 - p)$ will be close to 1 and so $\sigma^2 \approx np = \lambda$ and you saw (in Section 2.2) that for a Poisson variable $E(X) = Var(X)$. However an approximation can be used if p is close to 1 by using the complementary random variable (counting the number of failures instead of the number of successes).

Example 8

A garden centre states that 95% of its daffodil bulbs will produce flowers the following season. Gill buys 100 bulbs. Use a suitable approximation to estimate the probability that at least 96 of the bulbs will flower next season.

Let X = the number of bulbs that flower
$X \sim B(100, 0.95)$

Let Y = the number of bulbs that do not flower
$Y \sim B(100, 0.05)$
Using a Poisson approximation
$Y \approx \sim Po(5)$
$P(X \geqslant 96) = P(Y \leqslant 4)$
$= 0.4405$

Define a suitable random variable.

n is large but p is close to 1 so a Poisson approximation cannot be used.

Define the complementary variable.

Now n is large, p is small and $np = 5 < 10$ so a Poisson approximation can be used.

At least 96 flower means $X \geqslant 96$ and this means that 4 or fewer don't so $Y \leqslant 4$.

Use tables with $\lambda = 5$ and $x = 4$.

Some calculators can find the binomial probability (0.43598...) very easily without the need to use an approximation or the complementary variable. This can provide a useful check in the examination but if the question tells you to use a suitable approximation you will gain no credit for using a calculator to find the binomial probability.

Exercise 2D

1 The random variable $X \sim B(80, 0.10)$. Using a suitable approximation, find
a $P(X \geqslant 1)$, **b** $P(X \leqslant 6)$.

2 The random variable $X \sim B(120, 0.02)$. Using a suitable approximation, find
a $P(X = 1)$, **b** $P(X \geqslant 3)$.

3 The random variable $X \sim B(50, 0.05)$. Find the percentage error in $P(X \leqslant 4)$ when X is approximated by a Poisson distribution.

Resources

28/29 of 157

Unique Exam café feature:

- Relax and prepare – revision planner; hints and tips; common mistakes
- Refresh your memory – revision checklist; language of the examination; glossary
- Get the result! – fully worked examination-style paper with chief examiner's commentary

Solutionbank Edexcel AS and A Level Modular Mathematics

Statistics 2

1 Binomial distribution

Exercises: A Questions: 1

Question:

A large bag contains counters of different colours.
Find the number of arrangements for the following selections

a 5 counters all of different colours,

b 5 counters where 3 are red and 2 are blue,

c 7 counters where 2 are red and 5 are green,

d 10 counters where 4 are blue and 6 are yellow,

e 20 counters where 2 are yellow and 18 are black.

Solution:

a $5! = 120$

Question Hint Solution Print Help

edexcel

Solutionbank

- Hints and solutions to every question in the textbook
- Solutions and commentary for all review exercises and the practice examination paper

Published by Pearson Education Limited, a company incorporated in England and Wales, having its registered office at Edinburgh Gate, Harlow, Essex, CM20 2JE. Registered company number: 872828

Edexcel is a registered trademark of Edexcel Limited

14 13 12
10 9 8

British Library Cataloguing in Publication Data is available from the British Library on request.

ISBN 978 0 435519 13 1

Edited by Susan Gardner
Typeset by Tech-Set Ltd
Illustrated by Tech-Set Ltd
Index by Indexing Specialists (UK) Ltd
Cover design by Christopher Howson
Picture research by Chrissie Martin
Cover photo/illustration © Science Photo Library / Laguna Design
Printed in China (CTPS/08)

Acknowledgements
The author and publisher would like to thank the following individuals and organisations for permission to reproduce photographs:

Constructionphotography.com / Giles Barnard p**1**; Alamy Images / John Bower Pollution p**19**; Corbis / Helen King p**37**; Alamy Images / Michael Arthur Thompson p**81**; iStockPhoto.com / Daniel Gilbey p**91**; Alamy Images / Ace Stock Ltd p**105**

Every effort has been made to contact copyright holders of material reproduced in this book. Any omissions will be rectified in subsequent printings if notice is given to the publishers.

After studying this chapter you should

- know when a binomial distribution is a suitable model for a practical situation
- know how to calculate probabilities using a binomial distribution
- know how to use tables to find cumulative binomial probabilities.

1 Binomial distribution

From rolling dice to quality control on a manufacturing production line, the binomial distribution has proved to be a very useful model to describe a wide range of problems.

1.1 Using factorial notation to find the number of arrangements of some objects.

In S1 you saw how to list the number of different arrangements of some objects.

Example 1

Find all possible arrangements of

a 3 objects where one is red, one is blue and one is green,

b 4 objects where 2 are red and 2 are blue.

a The first object can be chosen in 3 ways.
The second object can be chosen in 2 ways.
The third object can be chosen in only 1 way.

Suppose the first one is red, then there are 2 possibilities for the second. If the first was blue then there are still 2 possibilities for the second. So the number of ways should be multiplied.

Here there are $3 \times 2 \times 1 = 6$ possible arrangements.

$3 \times 2 \times 1$ can be written as 3! or 3 factorial.

These can be listed as follows:

RBG BRG GRB
RGB BGR GBR

Try and list the possible ways in a systematic manner. For example, suppose the first is R then…

b If the red objects are labelled R_1 and R_2 and the blue objects B_1 and B_2, then you can treat the objects as 4 different ones and there are 4! ways of arranging them. However, arrangements with $R_1R_2\ldots$ are identical to arrangements with $R_2R_1\ldots$ and so the total number needs to be divided by 2.

Dividing by 2 is the same as dividing by 2!

A similar argument applies to the Bs and so number of arrangements is: $\dfrac{4!}{2! \times 2!} = 6$

You divide by:
2! for the 2 reds and 2! for the 2 blues.

These can be listed as follows:

RRBB RBBR RBRB
BBRR BRRB BRBR

Try and work systematically when listing. For example here the second row is simply the first with R swapped for B.

- n different objects can be arranged in $n! = n \times (n-1) \times (n-2) \ldots 3 \times 2 \times 1$ ways.
- n objects with a of one type and $(n-a)$ of another can be arranged in $\frac{n!}{a! \times (n-a)!}$ ways.

This is sometimes written as $\binom{n}{a}$ or nCa. You met this formula in C2 in connection with the binomial theorem.

Example 2

A child is playing with some identically shaped coloured bricks. Find the number of different arrangements that can be made from

a 6 red and 3 blue bricks, **b** 5 red and 7 blue bricks.

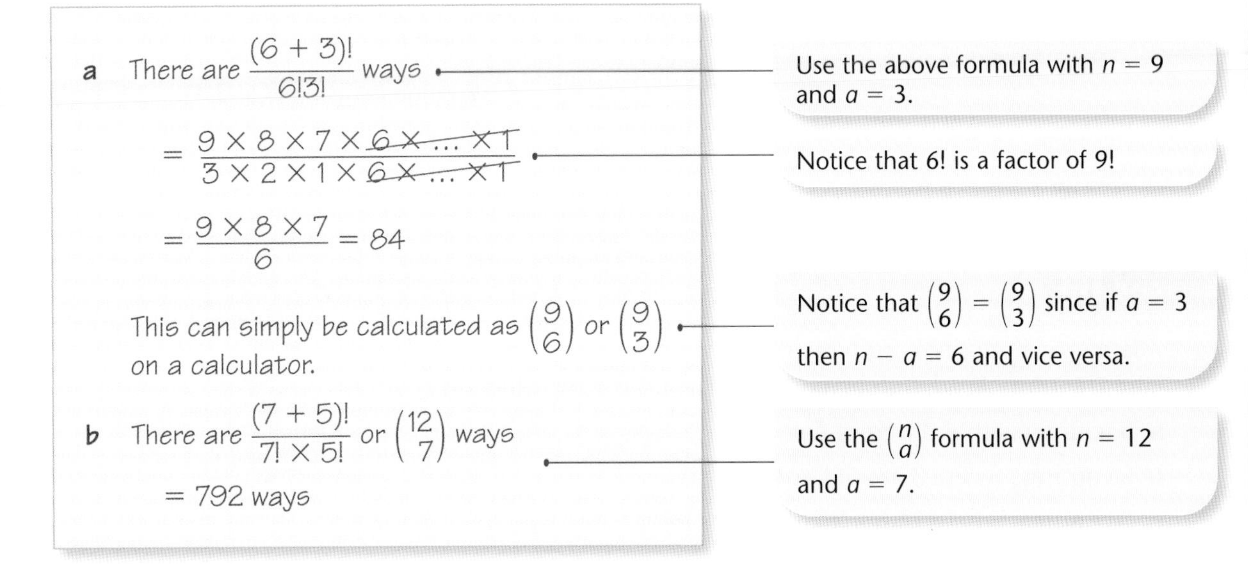

a There are $\frac{(6+3)!}{6!3!}$ ways

$= \frac{9 \times 8 \times 7 \times 6 \times \ldots \times 1}{3 \times 2 \times 1 \times 6 \times \ldots \times 1}$

$= \frac{9 \times 8 \times 7}{6} = 84$

This can simply be calculated as $\binom{9}{6}$ or $\binom{9}{3}$ on a calculator.

b There are $\frac{(7+5)!}{7! \times 5!}$ or $\binom{12}{7}$ ways

$= 792$ ways

Use the above formula with $n = 9$ and $a = 3$.

Notice that 6! is a factor of 9!

Notice that $\binom{9}{6} = \binom{9}{3}$ since if $a = 3$ then $n - a = 6$ and vice versa.

Use the $\binom{n}{a}$ formula with $n = 12$ and $a = 7$.

The formula for the number of arrangements can be helpful in probability questions.

Example 3

A fair die is rolled 8 times. Find the probability of:

a no sixes, **b** only 3 sixes, **c** four twos and 4 sixes.

Let S = the outcome 'the die lands on a six'.

a $P(\text{no sixes}) = P(S'\,S'\,S'\,S'\,S'\,S'\,S'\,S')$

$= \left(\frac{5}{6}\right)^8 = 0.23256\ldots$

$= 0.233$ (3 s.f.)

So S' is the outcome that the die does not land on a six.

There is only one arrangement of 8 rolls each having the outcome S'.

$P(S) = \frac{1}{6}$ and so $P(S') = 1 - \frac{1}{6} = \frac{5}{6}$.

b $P(3 \text{ sixes}) = P(3S \text{ and } 5S' \text{ in any order})$

$= \left(\frac{1}{6}\right)^3\left(\frac{5}{6}\right)^5 \times \frac{8!}{3!5!}$

$= 0.10419\ldots$

$= 0.104$ (3 s.f.)

c Let T = the outcome 'the die lands on a two'.

$P(4T \text{ and } 4S \text{ in any order})$

$= \left(\frac{1}{6}\right)^4\left(\frac{1}{6}\right)^4 \times \frac{8!}{4!4!}$

$= 0.00004167\ldots$

$= 0.0000417$ (3 s.f.)

$\left(\frac{1}{6}\right)^3$ for the 3 sixes.

$\left(\frac{5}{6}\right)^5$ for the 5 non-sixes.

There are $\binom{8}{3}$ or $\binom{8}{5} = \frac{8!}{5!3!}$ arrangements of $3S$ and $5S'$.

There are $4S$ and $4T$ and $P(S) = P(T) = \frac{1}{6}$.

$\left(\frac{1}{6}\right)^4$ for $4T$ and $\left(\frac{1}{6}\right)^4$ for $4S$.

There are $\binom{8}{4} = \frac{8!}{4!4!}$ arrangements of $4S$ and $4T$.

Exercise 1A

1 A large bag contains counters of different colours.
Find the number of arrangements for the following selections

a 5 counters all of different colours,

b 5 counters where 3 are red and 2 are blue,

c 7 counters where 2 are red and 5 are green,

d 10 counters where 4 are blue and 6 are yellow,

e 20 counters where 2 are yellow and 18 are black.

2 A bag contains 4 red, 3 green and 8 yellow beads.
Five beads are selected at random from the bag without replacement.
Find the probability that they are

a 5 yellow beads,

b 2 red and 3 yellow,

c 4 red and 1 green.

3 A fair die is rolled 7 times.
Find the probability of getting

a no fives,

b only 3 fives,

c 4 fives and 3 sixes.

1.2 Using the binomial theorem to find probabilities.

Example 4

A biased coin has probability p of landing heads when it is thrown.
The coin is thrown 10 times.
Find, in terms of p, the probability of the coin landing heads

a 10 times, **b** 7 times, **c** 4 times.

Let the random variable X = number of times (out of a possible 10) the coin lands heads.

a $P(X = 10) = p \times p \times p \times p \times p \times p \times p \times p \times p \times p = p^{10}$

b If the coin does not land heads it must land tails.
Let q = probability of landing tails

$P(X = 7) = \binom{10}{7} p^7 q^3$

$= 120p^7q^3$

$= 120p^7(1 - p)^3$

c $P(X = 4) = \binom{10}{4} p^4 q^6$

$= 210p^4q^6$

$= 210p^4(1 - p)^6$

There is only 1 arrangement of 10 heads.

Assuming each throw is independent you can multiply the probabilities together.

There are only 2 possibilities, H or T.

$q = 1 - p$

$\binom{10}{7} = 120$ for the number of arrangements of $7H$ and $3T$. Then p^7 for the 7 heads and q^3 for the 3 tails.

Don't forget to replace q with $(1 - p)$ as the question asks for the answer in terms of p.

$\binom{10}{4} = 210$ for the number of arrangements of $4H$ and $6T$. Then p^4 for the $4H$ and q^6 for the $6T$.

You met the **binomial expansion** in C2. Consider the binomial expansion of:

$$(p + q)^{10} = \underline{p^{10}} + \binom{10}{9} p^9 q + \binom{10}{8} p^8 q^2 + \underline{\binom{10}{7} p^7 q^3} + \ldots + \underline{\binom{10}{4} p^4 q^6} + \ldots + q^{10}$$

You should notice that the probabilities in Example 4 are terms in this binomial expansion. You could use the full binomial expansion to write down the probability distribution for the random variable X.

x	0	1	2	3	4	5	6	7	8	9	10
$P(X = x)$	q^{10}	…	…	…	$\binom{10}{4} p^4 q^6$	…	…	$\binom{10}{7} p^7 q^3$	…	…	p^{10}

Notice that the power of p is the value of x.

1.3 When a binomial distribution is a suitable model.

If you look again at Example 4 you can see that there are 4 conditions required for a binomial distribution.

- **A fixed number of trials, n.**

The coin was thrown 10 times, so $n = 10$.

- **Each trial should be success or failure.**

The coin could land H or T on each throw.

The distribution is called **bi**nomial because there are only 2 cases for each trial.

- **The trials are independent.**

You use this when you multiply the probabilities together.

- **The probability of success, p, at each trial is constant.**

You assumed that the probability of a coin landing heads was p for each throw.

- **If these conditions are satisfied we say that the random variable X(= the number of successes in n trials) has a binomial distribution and write**

  $X \sim \mathrm{B}(n, p)$

'B' for binomial
n for the number of trials
p for the probability of success at each trial.

  and $\mathrm{P}(X = x) = \binom{n}{x} p^x (1 - p)^{n-x}$

This formula is in the formula book.

Sometimes n is called the index and p the parameter of the binomial distribution.

Example 5

The random variable $X \sim \mathrm{B}(12, \frac{1}{6})$

Find **a** $\mathrm{P}(X = 2)$ **b** $\mathrm{P}(X = 9)$ **c** $\mathrm{P}(X \leqslant 1)$

a $\mathrm{P}(X = 2) = \binom{12}{2}\left(\frac{1}{6}\right)^2\left(\frac{5}{6}\right)^{10} = \frac{12!}{2!10!}\left(\frac{1}{6}\right)^2\left(\frac{5}{6}\right)^{10}$

$= 0.29609\ldots$

$= 0.296$ (3 s.f.)

Use the formula with $n = 12$, $p = \frac{1}{6}$ and $x = 2$.

b $\mathrm{P}(X = 9) = \binom{12}{9}\left(\frac{1}{6}\right)^9\left(\frac{5}{6}\right)^3$

$= 0.00001263\ldots$

$= 0.0000126$ (3 s.f.)

Use the formula with $n = 12$, $p = \frac{1}{6}$ and $x = 9$.

c $P(X \leqslant 1) = P(X = 0) + P(X = 1)$

$= \left(\frac{5}{6}\right)^{12} + \binom{12}{1}\left(\frac{1}{6}\right)^{1}\left(\frac{5}{6}\right)^{11}$

$= 0.112156\ldots + 0.26917\ldots$

$= 0.38133\ldots$

$= 0.381$ (3 s.f.)

Interpret the inequality to identify the values of X and then use the formula.

Example 6

Explain whether or not a binomial distribution can be used to model the following situations. In cases where it can be used, give a definition of the random variable and suggest suitable values for n and p.

a The number of throws of a die until a six is obtained.

b The number of girls in a family of 4 children.

c The number of red balls selected when 3 balls are drawn from an urn which contains 15 white and 5 red balls.

a Binomial is *not* suitable since the number of throws (trials) is not fixed.

b Number of trials is fixed ($n = 4$).
Each trial is success or failure.
Trials are probably independent.
The probability of having a girl should be constant.
So X = number of girls in families of 4 children and $X \sim B(4, 0.5)$

c Let R = number of red balls selected.
The number of trials is fixed ($n = 3$)
Each trial is success (red) or failure (white).
If the balls are drawn *with replacement* the probability of selecting a red is constant and $p = \frac{1}{4}$.
The trials will be independent too (if the first is red it does not alter the probability of the second being red).
So $R \sim B(3, \frac{1}{4})$

In questions of this type consider carefully whether any of the four conditions for a binomial distribution are not satisfied. Remember *all* four conditions *must* hold for a binomial.

A 'trial' is having a child.

Having a girl is a 'success', a boy is 'failure'.

The gender of the first child shouldn't influence the gender of the second. (Identical twins may be a problem here.)

It is not clear what p should be. Do you take a national average? However $p = 0.5$ will probably be a good starting value for your model.

The number of balls to be drawn is fixed at 3.

Selecting *with replacement* is critical here for R to have a binomial distribution.

If the balls are selected *without* replacement then p is not constant, so R would not be from a binomial distribution.

Exercise 1B

1 The random variable $X \sim B(8, \frac{1}{3})$. Find

a $P(X = 2)$, **b** $P(X = 5)$, **c** $P(X \leqslant 1)$.

2 The random variable $Y \sim B(6, \frac{1}{4})$. Find

a $P(Y = 3)$, **b** $P(Y = 1)$, **c** $P(Y \geqslant 5)$.

3 The random variable $T \sim B(15, \frac{2}{3})$. Find

a $P(T = 5)$, **b** $P(T = 10)$, **c** $P(3 \leqslant T \leqslant 4)$.

4 A balloon manufacturer claims that 95% of his balloons will not burst when blown up. If you have 20 of these balloons to blow up for a birthday party

a what is the probability that none of them burst when blown up?

b Find the probability that exactly 2 balloons burst.

5 A student suggests using a binomial distribution to model the following situations. Give a description of the random variable, state any assumptions that must be made and give possible values for n and p.

a A sample of 20 bolts is checked for defects from a large batch. The production process should produce 1% of defective bolts.

b Some traffic lights have three phases: stop 48% of the time, wait or get ready 4% of the time and go 48% of the time. Assuming that you only cross a traffic light when it is in the go position, model the number of times that you have to wait or stop on a journey passing through 6 sets of traffic lights.

c When Stephanie plays tennis with Timothy on average one in eight of her serves is an 'ace'. How many 'aces' does Stephanie serve in the next 30 serves against Timothy?

6 State which of the following can be modelled with a binomial distribution and which can not. Give reasons for your answers.

a Given that 15% of people have blood that is Rhesus negative (Rh^-), model the number of pupils in a statistics class of 14 who are Rh^-.

b You are given a fair coin and told to keep tossing it until you obtain 4 heads in succession. Model the number of tosses you need.

c A certain car manufacturer produces 12% of new cars in the colour red, 8% in blue, 15% in white and the rest in other colours. You make a note of the colour of the first 15 new cars of this make. Model the number of red cars you observe.

7 A fair die is rolled repeatedly. Find the probability that

a the first 6 occurs on the fourth roll,

b there are 3 sixes in the first 10 rolls.

8 A coin is biased so that the probability of it landing on heads is $\frac{2}{3}$. The coin is tossed repeatedly. Find the probability that

a the first tail will occur on the fifth toss,

b in the first 7 tosses there will be exactly 2 tails.

1.4 Using the tables of the cumulative distribution function of the binomial distribution to find probabilities.

In S1 you met the cumulative distribution function $F(x) = P(X \leqslant x)$.

For the binomial distribution $X \sim B(n, p)$ there are tables giving $P(X \leqslant x)$ for various values of n and p. Using these tables can make potentially lengthy calculations much shorter.

Example 7

The random variable $X \sim B(20, 0.4)$. Find

a $P(X \leqslant 7)$, **b** $P(X < 6)$, **c** $P(X \geqslant 15)$.

a $P(X \leqslant 7) = 0.4159$

Use $n = 20$, $p = 0.4$ and $x = 7$ in the cumulative binomial distribution tables.

Always quote values from the tables in full.

b $P(X < 6) = P(X \leqslant 5)$

$= 0.1256$

Since X is a discrete random variable, $X < 6$ means $X \leqslant 5$.

The tables give $P(X \leqslant x)$ so use $x = 5$.

c $P(X \geqslant 15) = 1 - P(X \leqslant 14)$

$= 1 - 0.9984$

$= 0.0016$

$P(X \geqslant 15) + P(X \leqslant 14) = 1$ (since the total probability $= 1$). The tables only give $P(X \leqslant x)$ so you must write $P(X \geqslant 15)$ as $1 - P(X \leqslant 14)$.

The tables can also be used to find $P(X = x)$ since $P(X = x) = P(X \leqslant x) - P(X \leqslant (x - 1))$.

Example 8

The random variable $X \sim B(25, 0.25)$. Find

a $P(X \leqslant 6)$, **b** $P(X = 6)$, **c** $P(X > 20)$, **d** $P(6 < X \leqslant 10)$.

a $P(X \leqslant 6) = 0.5611$

Use cumulative binomial tables with $n = 25$, $p = 0.25$ and $x = 6$.

b $P(X = 6) = P(X \leqslant 6) - P(X \leqslant 5)$

$= 0.5611 - 0.3783$

$= 0.1828$

Write $P(X = 6)$ in terms of cumulative probabilities.

Use tables.

c $P(X > 13) = 1 - P(X \leqslant 13)$

$= 1 - 0.9991$

$= 0.0009$

The tables give $P(X \leqslant x)$ so you need to remember $P(X > 13) = P(X \geqslant 14)$ and use the fact that $P(X \geqslant 14) + P(X \leqslant 13) = 1$. Then use tables.

d $P(6 < X \leqslant 10) = P(X = 7, 8, 9 \text{ or } 10)$

$= P(X \leqslant 10) - P(X \leqslant 6)$

$= 0.9703 - 0.5611$

$= 0.4092$

Write the probability as the difference of two cumulative probabilities.

Use tables.

Sometimes questions are set in context and there are many different forms of words that can be used to ask for probabilities. The correct interpretation of these phrases is critical, especially when dealing with discrete distributions such as the binomial and Poisson (see Chapter 2). The table gives some examples.

Phrase	Means	To use tables…
… greater than 5 …	$X > 5$	$1 - P(X \leqslant 5)$
… no more than 3 …	$X \leqslant 3$	$P(X \leqslant 3)$
… at least 7 …	$X \geqslant 7$	$1 - P(X \leqslant 6)$
… fewer than 10 …	$X < 10$	$P(X \leqslant 9)$
… at most 8 …	$X \leqslant 8$	$P(X \leqslant 8)$

Example 9

A spinner is designed so that the probability it lands on red is 0.3. Jane has 12 spins. Find the probability that Jane obtains

a no more than 2 reds, **b** at least 5 reds.

Jane decides to use this spinner for a class competition. She wants the probability of winning a prize to be < 0.05. Each member of the class will have 12 spins and the number of reds will be recorded.

c Find how many reds are needed to win a prize.

Let X = the number of reds in 12 spins.

$X \sim B(12, 0.3)$

Define a suitable random variable. This makes it easier to rewrite the question in terms of probabilities and it can help you determine the distribution.

a $P(X \leqslant 2) = 0.2528$

… no more than 2 means $X \leqslant 2$.

Use tables with $n = 12$, $p = 0.3$ and $x = 2$.

b $P(X \geqslant 5)$

$= 1 - P(X \leqslant 4)$

$= 1 - 0.7237$

$= 0.2763$

… at least 5 means $X \geqslant 5$.

Since tables only give $P(X \leqslant x)$ you need to write as $1 - P(X \leqslant 4)$ and then use tables with $n = 12$, $p = 0.3$ and $x = 4$.

c Let r = the smallest number of reds needed to win a prize.

Require: $P(X \geqslant r) < 0.05$

Form a probability statement to represent the condition for winning a prize.

From tables:

$P(X \leqslant 5) = 0.8822$

$P(X \leqslant 6) = 0.9614$

$P(X \leqslant 7) = 0.9905$

Use the tables with $n = 12$ and $p = 0.3$. You want values of x that give cumulative probabilities > 0.95 (since $0.95 = 1 - 0.05$).

So: $P(X \leqslant 6) = 0.9614$ implies that

$P(X \geqslant 7) = 1 - 0.9614$

$= 0.0386 < 0.05$

Since $x = 6$ gives the first value > 0.95 use this probability and find $r = 7$.

You should show that you are comparing this with the given condition.

So 7 or more reds will win a prize.

Always make sure that your final answer is related back to the context of the original question.

This technique of using the tables backwards as in part **c** will be met again in connection with critical regions for hypothesis tests in Chapter 7.

Exercise 1C

1 The random variable $X \sim B(9, 0.2)$. Find

a $P(X \leqslant 4)$, **b** $P(X < 3)$, **c** $P(X \geqslant 2)$, **d** $P(X = 1)$.

2 The random variable $X \sim B(20, 0.35)$. Find

a $P(X \leqslant 10)$, **b** $P(X > 6)$, **c** $P(X = 5)$, **d** $P(2 \leqslant X \leqslant 7)$.

3 The random variable $X \sim B(40, 0.45)$. Find

a $P(X < 20)$, **b** $P(X > 16)$, **c** $P(11 \leqslant X \leqslant 15)$, **d** $P(10 < X < 17)$.

4 The random variable $X \sim B(30, 0.15)$. Find

a $P(X > 8)$, **b** $P(X \leqslant 4)$, **c** $P(2 \leqslant X < 10)$, **d** $P(X = 4)$.

5 Eight fair coins are tossed and the total number of heads showing is recorded. Find the probability of

a no heads, **b** at least 2 heads, **c** more heads than tails.

6 For a particular type of plant 25% have blue flowers. A garden centre sells these plants in trays of 15 plants of mixed colours. A tray is selected at random.
Find the probability that the number of blue flowers this tray contains is

a exactly 4, **b** at most 3, **c** between 3 and 6 (inclusive).

7 The random variable $X \sim B(50, 0.40)$. Find

a the largest value of k such that $P(X \leqslant k) < 0.05$,

b the smallest number r such that $P(X > r) < 0.01$.

8 The random variable $X \sim B(40, 0.10)$. Find

a the largest value of k such that $P(X < k) < 0.02$,

b the smallest number r such that $P(X > r) < 0.01$,

c $P(k \leqslant X \leqslant r)$.

9 In a town, 30% of residents listen to the local radio.
Ten residents are chosen at random.

a State the distribution of the random variable

X = the number of these 10 residents that listen to the local radio.

b Find the probability that at least half of these 10 residents listen to local radio.

c Find the smallest value of s so that $P(X \geqslant s) < 0.01$.

10 A factory produces a component for the motor trade and 5% of the components are defective. A quality control officer regularly inspects a random sample of 50 components. Find the probability that the next sample contains

a fewer than 2 defectives,

b more than 5 defectives.

The officer will stop production if the number of defectives in the sample is greater than a certain value d. Given that the officer stops production less than 5% of the time,

c find the smallest value of d.

1.5 You can use simple formulae to find the mean and variance of the binomial distribution.

If $X \sim B(n, p)$ then

$$E(X) = \mu = np$$
$$\text{Var}(X) = \sigma^2 = np(1 - p)$$

These formulae are given in the formula booklet. You are not expected to know how to prove these formulae in the S2 exam.

Example 10

A fair, 4-sided die has the numbers 1, 2, 3 and 4 on its faces. The die is rolled 20 times. The random variable X represents the number of 4s obtained.

a Find the mean and variance of X.

b Find $\mathrm{P}(X < \mu - \sigma)$.

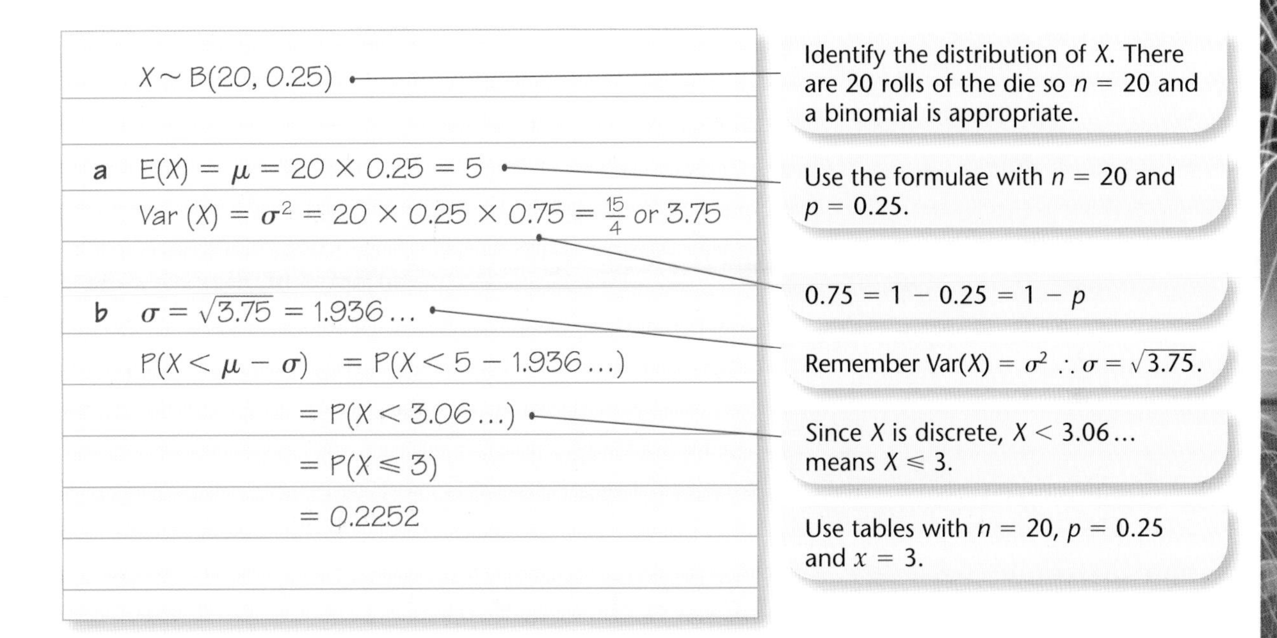

$X \sim \mathrm{B}(20, 0.25)$

Identify the distribution of X. There are 20 rolls of the die so $n = 20$ and a binomial is appropriate.

a $\mathrm{E}(X) = \mu = 20 \times 0.25 = 5$

$\mathrm{Var}(X) = \sigma^2 = 20 \times 0.25 \times 0.75 = \frac{15}{4}$ or 3.75

Use the formulae with $n = 20$ and $p = 0.25$.

$0.75 = 1 - 0.25 = 1 - p$

b $\sigma = \sqrt{3.75} = 1.936\ldots$

Remember $\mathrm{Var}(X) = \sigma^2 \therefore \sigma = \sqrt{3.75}$.

$\mathrm{P}(X < \mu - \sigma) = \mathrm{P}(X < 5 - 1.936\ldots)$

$= \mathrm{P}(X < 3.06\ldots)$

Since X is discrete, $X < 3.06\ldots$ means $X \leqslant 3$.

$= \mathrm{P}(X \leqslant 3)$

$= 0.2252$

Use tables with $n = 20$, $p = 0.25$ and $x = 3$.

Example 11

David believes that 35% of people in a certain town will vote for him in the next election and he commissions a survey to verify this. Find the minimum number of people the survey should ask to have a mean number of more than 100 voting for David.

Let n = the number of people asked.

Let X = the number of people (out of n) voting for David.

Define a suitable random variable. The proportion of 35% will give $p = 0.35$.

$X \sim \mathrm{B}(n, 0.35)$

$\mathrm{E}(X) = 0.35n$

Use the formula $\mu = np$.

So $0.35n > 100$

$$n > \frac{100}{0.35} = 285.7\ldots$$

Form a suitable inequality, using the fact that the mean is greater than 100, and solve to find n.

So ($n =$) 286 people should be asked.

Example 12

An examiner is trying to design a multiple choice test. For students answering the test at random, he requires that the mean score on the test should be 20 and the standard deviation should be at least 4. Assuming that each question has the same number of alternative answers, find how many questions and how many alternative answers each question should have. The number of alternatives for each question should be as few as possible.

Let n = the number of questions

and p = the probability of guessing a correct answer.

Let X = the number of questions answered correctly at random

$X \sim B(n, p)$

Define n and p to form a suitable binomial distribution.

$E(X) = 20$ so $np = 20$

St. dev $\geqslant 4$ so $np(1 - p) \geqslant 4^2$

Use the formulae for mean and variance. Remember variance = (st. dev)^2.

So $20(1 - p) \geqslant 16$

$1 - p \geqslant 0.8$

$0.2 \geqslant p$

Substitute $np = 20$ and solve for p.

So the examiner should use $p = 0.2$, i.e. have 5 alternatives for each question.

$np = 20$ with $p = 0.2$ gives $n = 100$.

So the test should have 100 questions and 5 alternative answers for each.

Exercise 1D

1 A fair cubical die is rolled 36 times and the random variable X represents the number of sixes obtained. Find the mean and variance of X.

2 **a** Find the mean and variance of the random variable $X \sim B(12, 0.25)$.
b Find $P(\mu - \sigma < X < \mu + \sigma)$.

3 **a** Find the mean and variance of the random variable $X \sim B(30, 0.40)$.
b Find $P(\mu - \sigma < X \leqslant \mu)$.

4 It is estimated that 1 in 20 people are left-handed.
a What size sample should be taken to ensure that the expected number of left-handed people in the sample is 3?
b What is the standard deviation of the number of left-handed people in this case?

5 An experiment is conducted with a fair die to examine the number of sixes that occur. It is required to have the standard deviation smaller than 1. What is the largest number of throws that can be made?

6 The random variable $X \sim B(n, p)$ has a mean of 45 and standard deviation of 6. Find the value of n and the value of p.

1.6 Solving harder problems.

Example 13

In Joe's café 70% of customers buy a cup of tea.

a In a random sample of 20 customers find the probability that more than 15 buy a cup of tea.

The proportion of customers who buy a chocolate muffin at Joe's café is 0.35.

b Find the probability that Joe sells the first chocolate muffin to his fifth customer.

Assuming that buying a chocolate muffin and buying a cup of tea are independent events

c find the probability that in a random sample of 20 customers more than 15 buy a cup of tea and at least 6 buy a chocolate muffin.

a Let X = the number out of 20 who buy a cup of tea.

$X \sim B(20, 0.7)$

Since 0.7 is not in the tables you will need to consider the complementary random variable Y.

Let Y = the number of customers who do not buy a cup of tea.

$Y \sim B(20, 0.3)$

Note that $X > 15$ is the same as $Y \leqslant 4$ and then use tables with $n = 20$ and $p = 0.3$.

$P(X > 15) = P(Y \leqslant 4) = 0.2375$

b Let N = the number of the first customer who buys a chocolate muffin.

Define the random variable carefully. Note the variable is the number of the customer not the number of muffins.

N is *not* binomial since there is no fixed value for n.

If C represents the event that a person buys a chocolate muffin then the first 5 customers must be $C'C'C'C'C$

So $P(N = 5) = (0.65)^4 \times 0.35$

$= 0.062477\ldots$

$= 0.0625$ (3 s.f.)

There is only one case possible so no $\binom{n}{r}$ term.

c From part **a** $P(X > 15) = 0.2375$

Let M = number of customers out of 20 who buy a chocolate muffin

The probability for cups of tea was found in **a**. So define the variable M.

$M \sim B(20, 0.35)$

$P(M \geqslant 6) = 1 - P(M \leqslant 5)$

$= 1 - 0.2454$

$= 0.7546$

Use $P(M \geqslant 6) + P(M \leqslant 5) = 1$ and then use tables with $n = 20$ and $p = 0.35$.

So $P(M \geqslant 6 \text{ and } X > 15) = 0.7546 \times 0.2375$

$= 0.17921\ldots$

$= 0.179$ (3 s.f.)

To find the probability of both $X > 15$ and $M \geqslant 6$ you multiply the probabilities together since buying a chocolate muffin and a cup of tea are independent.

Example 14

Yummy sweets are sold in tubes of 15 sweets per tube. The sweets are of different flavours and 20% are blackcurrant.

a Show that the probability of a randomly chosen tube of Yummy sweets containing 3 or more blackcurrant ones is 0.602.

Robert buys 5 tubes of Yummy sweets.

b Find the probability that at least 4 of the tubes contain 3 or more blackcurrant sweets.

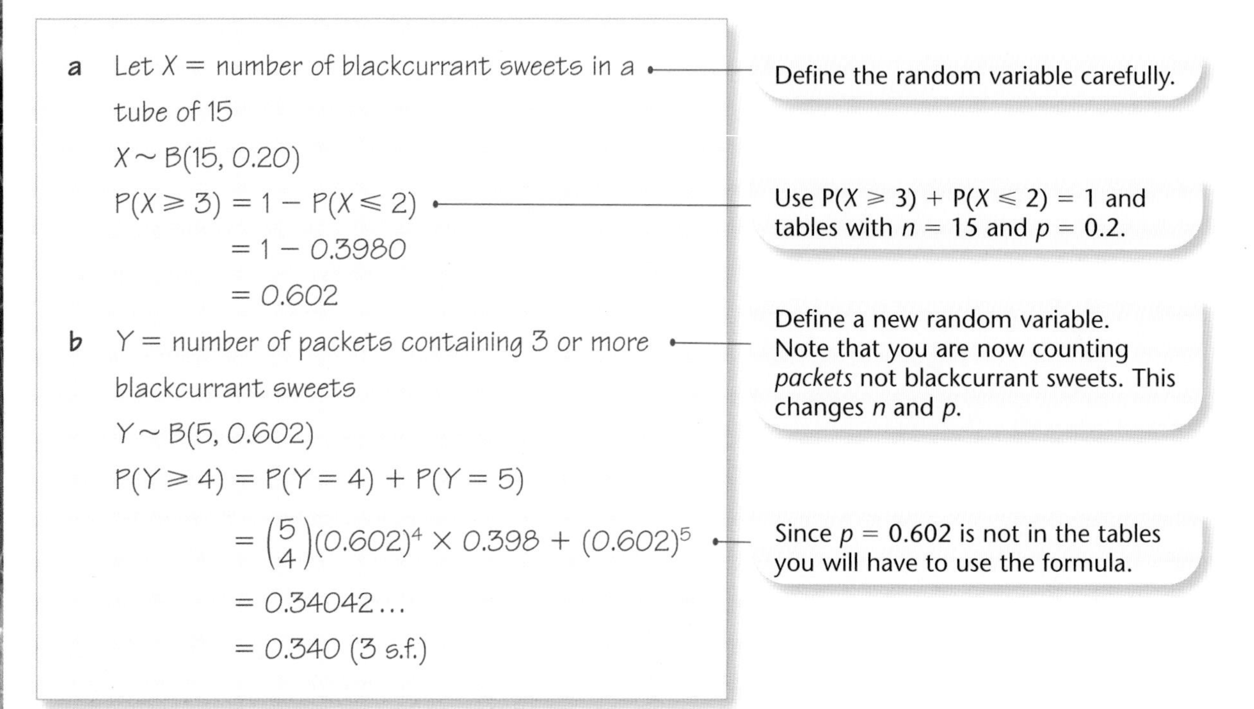

a Let X = number of blackcurrant sweets in a tube of 15

Define the random variable carefully.

$X \sim B(15, 0.20)$

$P(X \geqslant 3) = 1 - P(X \leqslant 2)$

Use $P(X \geqslant 3) + P(X \leqslant 2) = 1$ and tables with $n = 15$ and $p = 0.2$.

$= 1 - 0.3980$

$= 0.602$

b Y = number of packets containing 3 or more blackcurrant sweets

Define a new random variable. Note that you are now counting *packets* not blackcurrant sweets. This changes n and p.

$Y \sim B(5, 0.602)$

$P(Y \geqslant 4) = P(Y = 4) + P(Y = 5)$

$= \binom{5}{4}(0.602)^4 \times 0.398 + (0.602)^5$

Since $p = 0.602$ is not in the tables you will have to use the formula.

$= 0.34042\ldots$

$= 0.340$ (3 s.f.)

Mixed exercise 1E

1 A coin is biased so that the probability of a head is $\frac{2}{3}$. The coin is tossed repeatedly. Find the probability that

a the first tail will occur on the sixth toss,

b in the first 8 tosses there will be exactly 2 tails.

2 Records kept in a hospital show that 3 out of every 10 patients who visit the accident and emergency department have to wait more than half an hour. Find, to 3 decimal places, the probability that of the first 12 patients who come to the accident and emergency department

a none, **b** more than 2,

will have to wait more than half an hour.

3 A factory is considering two methods of checking the quality of production of the batches of items it produces.
Method I A random sample of 10 items is taken from a large batch and the batch is accepted if there are no defectives in this sample. If there are 2 or more defectives the batch is

rejected. If there is only 1 defective then another sample of 10 is taken and the batch is accepted if there are no defectives in this second sample, otherwise the whole batch is rejected.

Method II A random sample of 20 items is taken from a large batch and the batch is accepted if there is at most 1 defective in this sample, otherwise the whole batch is rejected.

The factory knows that 1% of items produced are defective and wishes to use the method of checking the quality of production for which the probability of accepting the whole batch is largest.

a Decide which method the factory should use.

b Determine the expected number of items sampled using Method I.

4 **a** State clearly the conditions under which it is appropriate to assume that a random variable has a binomial distribution.

A door-to-door canvasser tries to persuade people to have a certain type of double glazing installed. The probability that his canvassing at a house is successful is 0.05.

b Find the probability that he will have at least 2 successes out of the first 10 houses he canvasses.

c Find the number of houses he should canvass per day in order to average 3 successes per day.

d Calculate the least number of houses that he must canvass in order that the probability of his getting at least one success exceeds 0.99.

E

5 An archer fires arrows at a target and for each arrow, independently of all the others, the probability that it hits the bull's eye is $\frac{1}{8}$.

a Given that the archer fires 5 arrows, find the probability that fewer than 2 arrows hit the bull's eye.

The archer fires 5 arrows, collects them from a target and fires all 5 again.

b Find the probability that on both occasions fewer than 2 hit the bull's eye.

6 A completely unprepared student is given a true/false type test with 10 questions. Assuming that the student answers all the questions at random

a find the probability that the student gets all the answers correct.

It is decided that a pass will be awarded for 8 or more correct answers.

b Find the probability that the student passes the test.

7 A six-sided die is biased. When the die is thrown the number 5 is twice as likely to appear as any other number. All the other faces are equally likely to appear. The die is thrown repeatedly. Find the probability that

a the first 5 will occur on the sixth throw,

b in the first eight throws there will be exactly three 5s.

E

8 A manufacturer produces large quantities of plastic chairs. It is known from previous records that 15% of these chairs are green. A random sample of 10 chairs is taken.

a Define a suitable distribution to model the number of green chairs in this sample.

b Find the probability of at least 5 green chairs in this sample.

c Find the probability of exactly 2 green chairs in this sample.

9 A bag contains a large number of beads of which 45% are yellow. A random sample of 20 beads is taken from the bag. Use the binomial distribution to find the probability that the sample contains

a fewer than 12 yellow beads,

b exactly 12 yellow beads. *E*

Summary of key points

1 If $X \sim B(n, p)$ then

$$P(X = x) = \binom{n}{x} p^x (1 - p)^{n - x}$$

2 If $X \sim B(n, p)$ then

$$E(X) = np$$
$$\text{Var}(X) = np(1 - p)$$

3 Conditions for a binomial distribution

- A fixed number of trials, n.
- Each trial should be success or failure.
- The trials are independent.
- The probability of success, p, at each trial is constant.

After studying this chapter you should

- know when a Poisson distribution is a suitable model to use
- know how to calculate probabilities using a Poisson distribution
- know how to use tables of the cumulative frequency distribution of a Poisson
- know how and when to use the Poisson distribution as an approximation to the binomial distribution.

Poisson distribution

The Poisson distribution is named after the French mathematician Simeon Poisson. It has many applications in modern day life such as modelling radioactive decay, predicting the arrival of buses, describing the spread of trees in a forest and predicting the number of telephone calls arriving at an exchange.

2.1 Relating the exponential series and the Poisson distribution.

Your calculator will evaluate the exponential function e^x for various values of x
e.g. $e^{-2} = 0.1353\ldots$ and $e^{0.5} = 1.6487\ldots$
The function e^x can be defined as a series

$$e^x = 1 + \frac{x^1}{1!} + \frac{x^2}{2!} + \frac{x^3}{3!} + \ldots + \frac{x^r}{r!} + \ldots$$

Remember
$r! = r \times (r-1) \times (r-2) \times \ldots \times 2 \times 1$.

If you let $x = \lambda$ and remember that $\lambda^0 = 1$ this gives

$$e^\lambda = \lambda^0 + \frac{\lambda^1}{1!} + \frac{\lambda^2}{2!} + \frac{\lambda^3}{3!} + \ldots + \frac{\lambda^r}{r!} + \ldots$$

This series is infinite.

Dividing by e^λ gives

$$\frac{e^\lambda}{e^\lambda} = 1 = \lambda^0 e^{-\lambda} + \frac{\lambda^1 e^{-\lambda}}{1!} + \frac{\lambda^2 e^{-\lambda}}{2!} + \frac{\lambda^3 e^{-\lambda}}{3!} + \ldots + \frac{\lambda^r e^{-\lambda}}{r!} + \ldots$$

Remember $e^{-\lambda} = \frac{1}{e^\lambda}$.

Notice that the sum of the infinite series on the right-hand side equals 1 and so you could use these values as probabilities to define a probability distribution.

Remember to be a probability distribution the sum of the probabilities = 1.

Let X be a random variable, such that X takes the values 0, 1, 2, 3, ... then the probability distribution for X is

x	0	1	2	3	...	r	...
$P(X = x)$	$e^{-\lambda}$	$\frac{e^{-\lambda}\lambda}{1!}$	$\frac{e^{-\lambda}\lambda^2}{2!}$	$\frac{e^{-\lambda}\lambda^3}{3!}$	...	$\frac{e^{-\lambda}\lambda^r}{r!}$	...

And the probability function is

- $P(X = x) = \frac{e^{-\lambda}\lambda^x}{x!}$

This formula is given in the formula booklet.

- **We say that X has a Poisson distribution with parameter λ and write**

  $X \sim \text{Po}(\lambda)$

$\sim$ means 'is distributed'.
Po is for Poisson distribution.
λ is the parameter.

Example 1

The random variable $X \sim \text{Po}(1.2)$ find

a $P(X = 3)$, **b** $P(X \geqslant 1)$, **c** $P(3 < X \leqslant 5)$.

a $P(X = 3) = \frac{e^{-1.2} \times 1.2^3}{3!}$

$= 0.086743\ldots$

$= 0.0867$ (3 s.f.)

Use the formula with $\lambda = 1.2$ and $x = 3$.
Evaluate on a calculator and give the answer to 3 s.f.

b $P(X \geq 1) = 1 - P(X = 0)$

$= 1 - e^{-1.2}$

$= 1 - 0.30119\ldots$

$= 0.699$ (3 s.f.)

Since x can take any integer value ≥ 0 you need to use $P(X = 0) + P(X \geq 1) = 1$.

c $P(3 < X \leq 5) = P(X = 4) + P(X = 5)$

$= \frac{e^{-1.2} \times 1.2^4}{4!} + \frac{e^{-1.2} \times 1.2^5}{5!}$

$= 0.02602\ldots + 0.006245\ldots$

$= 0.0323$ (3 s.f.)

Interpret inequalities carefully since the Poisson is a *discrete* distribution.

Use the formula with $x = 4$ and $x = 5$.

Exercise 2A

1 The discrete random variable $X \sim \text{Po}(2.3)$. Find

a $P(X = 4)$, **b** $P(X \geq 1)$, **c** $P(4 < X < 6)$.

2 The discrete random variable $X \sim \text{Po}(5.7)$. Find

a $P(X = 7)$, **b** $P(X \leq 1)$, **c** $P(X > 2)$.

3 The random variable $Y \sim \text{Po}(0.35)$. Find

a $P(Y = 1)$, **b** $P(Y \geq 1)$, **c** $P(1 \leq Y < 3)$.

4 The random variable $X \sim \text{Po}(3.6)$. Find

a $P(X = 5)$, **b** $P(3 < X \leq 6)$, **c** $P(X < 2)$.

2.2 Using simple formulae to find the mean and variance of a Poisson distribution.

- **If the random variable $X \sim \text{Po}(\lambda)$ then it can be shown that**

  Mean of $X = \mu = \text{E}(X) = \lambda$

  Variance of $X = \sigma^2 = \text{Var}(X) = \lambda$

These formulae are given in the formula booklet.

- **The fact the mean equals the variance is an important property of a Poisson distribution and the presence or absence of this property can be a useful indicator of whether or not a Poisson distribution is a suitable model for a particular situation. The full conditions for a Poisson distribution are considered in Section 2.4.**

2.3 Using tables of the Poisson cumulative distribution function.

Example 2

The random variable $X \sim \text{Po}(2)$. Find

a $\text{P}(X \leqslant 4)$, **b** $\text{P}(X = 3)$, **c** $\text{P}(X > 2)$, **d** $\text{P}(3 \leqslant X < 7)$.

a $\text{P}(X) \leqslant 4) = 0.9473$

Use tables of the Poisson cumulative distribution function with $\lambda = 2$ and $x = 4$.

b $\text{P}(X = 3) = \text{P}(X \leqslant 3) - \text{P}(X \leqslant 2)$
$= 0.8571 - 0.6767$
$= 0.1804$

Write $\text{P}(X = 3)$ as a difference of two cumulative probabilities.

c $\text{P}(X > 2) = \text{P}(X \geqslant 3) = 1 - \text{P}(X \leqslant 2)$
$= 1 - 0.6767$
$= 0.3233$

As in Chapter 1 you need to use the fact that $\text{P}(X \geqslant 3) + \text{P}(X \leqslant 2) = 1$.

d $\text{P}(3 \leqslant X < 7) = \text{P}(3 \leqslant X \leqslant 6)$
$= \text{P}(X \leqslant 6) - \text{P}(X \leqslant 2)$
$= 0.9955 - 0.6767$
$= 0.3188$

Write the probability as a difference of two cumulative probabilities.

Example 3

The random variable $X \sim \text{Po}(7.5)$. Find the values of a, b and c such that

a $\text{P}(X \leqslant a) = 0.2414$, **b** $\text{P}(X < b) = 0.5246$, **c** $\text{P}(X \geqslant c) = 0.338$.

a $\text{P}(X \leqslant a) = 0.2414$
implies $a = 5$

Use tables with $\lambda = 7.5$. $\text{P}(X \leqslant 5) = 0.2414$.

b $\text{P}(X < b) = \text{P}(X \leqslant b - 1) = 0.5246$
so $b - 1 = 7$
$b = 8$

Use tables with $\lambda = 7.5$. $\text{P}(X \leqslant 7) = 0.5246$.

c $\text{P}(X \geqslant c) = 1 - \text{P}(X \leqslant c - 1) = 0.338$
so $\text{P}(X \leqslant c - 1) = 1 - 0.338$
$= 0.662$

Use tables with $\lambda = 7.5$. $\text{P}(X \leqslant 8) = 0.6620$.

so $c - 1 = 8$
$c = 9$

Example 4

The number of accidents on a stretch of motorway was monitored over a long period of time. The mean number of accidents per month was found to be 1.5 and the standard deviation was 1.2. An accident investigator suggests that the number of accidents per month, on this stretch of motorway, could be modelled by a Poisson distribution.

a Comment on this suggestion.

A Poisson distribution is used to model the number of accidents on this stretch of motorway in a randomly chosen month.

b Find the probability of no more than 2 accidents.

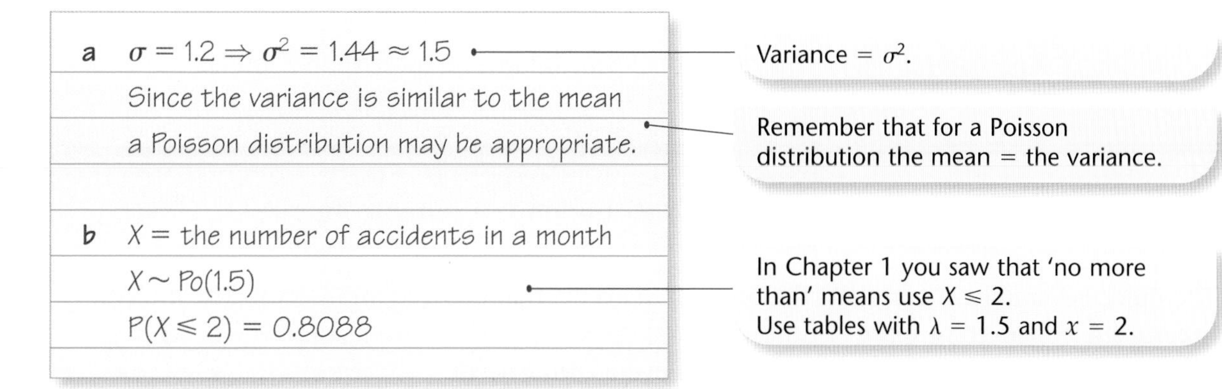

a $\sigma = 1.2 \Rightarrow \sigma^2 = 1.44 \approx 1.5$

Since the variance is similar to the mean a Poisson distribution may be appropriate.

b X = the number of accidents in a month

$X \sim \text{Po}(1.5)$

$\text{P}(X \leqslant 2) = 0.8088$

Variance = σ^2.

Remember that for a Poisson distribution the mean = the variance.

In Chapter 1 you saw that 'no more than' means use $X \leqslant 2$.
Use tables with $\lambda = 1.5$ and $x = 2$.

Exercise 2B

1 The random variable $X \sim \text{Po}(2.5)$. Find

a $\text{P}(X = 1)$, **b** $\text{P}(X > 2)$, **c** $\text{P}(X \leqslant 5)$, **d** $\text{P}(3 \leqslant X \leqslant 5)$.

2 The random variable $X \sim \text{Po}(6)$. Find

a $\text{P}(X \leqslant 3)$, **b** $\text{P}(X > 4)$, **c** $\text{P}(X = 5)$, **d** $\text{P}(2 < X \leqslant 7)$.

3 The random variable Y has a Poisson distribution with mean 4.5. Find

a $\text{P}(Y = 2)$, **b** $\text{P}(Y \leqslant 1)$, **c** $\text{P}(Y > 4)$, **d** $\text{P}(2 \leqslant Y \leqslant 6)$.

4 The random variable $X \sim \text{Po}(8)$. Find the values of a, b, c and d such that

a $\text{P}(X \leqslant a) = 0.3134$, **b** $\text{P}(X \leqslant b) = 0.7166$,

c $\text{P}(X < c) = 0.0996$, **d** $\text{P}(X > d) = 0.8088$.

5 The random variable $X \sim \text{Po}(3.5)$. Find the values of a, b, c and d such that

a $\text{P}(X \leqslant a) = 0.8576$, **b** $\text{P}(X > b) = 0.6792$,

c $\text{P}(X \leqslant c) \geqslant 0.95$, **d** $\text{P}(X > d) \leqslant 0.005$.

6 The number of telephone calls received at an exchange during a weekday morning follows a Poisson distribution with a mean of 6 calls per 5-minute period. Find the probability that

a there are no calls in the next 5 minutes,

b 3 or fewer calls are received in the next 5 minutes,

c fewer than 2 calls are received between 11:00 and 11:05,

d no more than 2 calls are received between 11:30 and 11:35.

7 The random variable $X \sim \text{Po}(9)$. Find

a $\mu = \text{E}(X)$,

b $\sigma =$ standard deviation of X,

c $\text{P}(\mu \leqslant X < \mu + \sigma)$,

d $\text{P}(X \leqslant \mu - \sigma)$.

8 The mean number of faults in $2\,\text{m}^2$ of cloth produced by a factory is 1.5.

a Find the probability of a $2\,\text{m}^2$ piece of cloth containing no faults.

b Find the probability that a $2\,\text{m}^2$ piece of cloth contains no more than 2 faults.

2.4 Deciding whether or not a Poisson distribution is a suitable model.

The random variable X represents the number of events that occur in an interval. The interval may be a fixed length in space or time.
If X is to have a Poisson distribution then the events must occur

- **singly** in space or time,
- **independently** of each other,
- **at a constant rate in the sense that the mean number of occurrences in the interval is proportional to the length of the interval.**

Such events are said to occur **randomly**.

You can use the constant rate idea to adjust the parameter of a Poisson distribution.

Example 5

Some river water contains on average 500 microbes per litre. A large bucket of the water is collected and after it has been well stirred a $1\,\text{cm}^3$ sample is examined.

a Explain the importance of stirring the water before taking the sample.

b Find the probability of there being no microbes in this sample.

c Find the probability of there being at least 4 microbes in the sample.

a The stirring avoids the possible problem of the microbes occurring in clusters. It should mean that they occur independently and singly. This is a condition for a Poisson distribution.

Relate the conditions for a Poisson distribution to the context of the question.

b λ = the average number of microbes in 1 cm^3

$= \frac{500}{1000} = 0.5$

X = the number of microbes in 1 cm^3

$P(X = 0) = e^{-\lambda} = e^{-0.5} = 0.6065$

c $P(X \geqslant 4) = 1 - P(X \leqslant 3)$

$= 1 - 0.9982$

$= 0.0018$

The microbes occur at a rate of 500 per litre. To use a Poisson distribution you need the average number in 1 cm^3.

Use the formula with $\lambda = 0.5$ and $X = 0$.

Use $P(X \geqslant 4) + P(X \leqslant 3) = 1$ and then tables with $\lambda = 0.5$ and $x = 3$.

Example 6

A shop sells radios at a rate of 2.5 per week.

a Find the probability that in a two-week period the shop sells at least 7 radios.

Deliveries of these radios come every 4 weeks.

b Find the probability of selling fewer than 12 radios in a four-week period.

The manager wishes to make sure that the probability of the shop running out of radios during a four-week period is less than 0.01.

c Find the smallest number of radios the manager should have in stock immediately after the delivery.

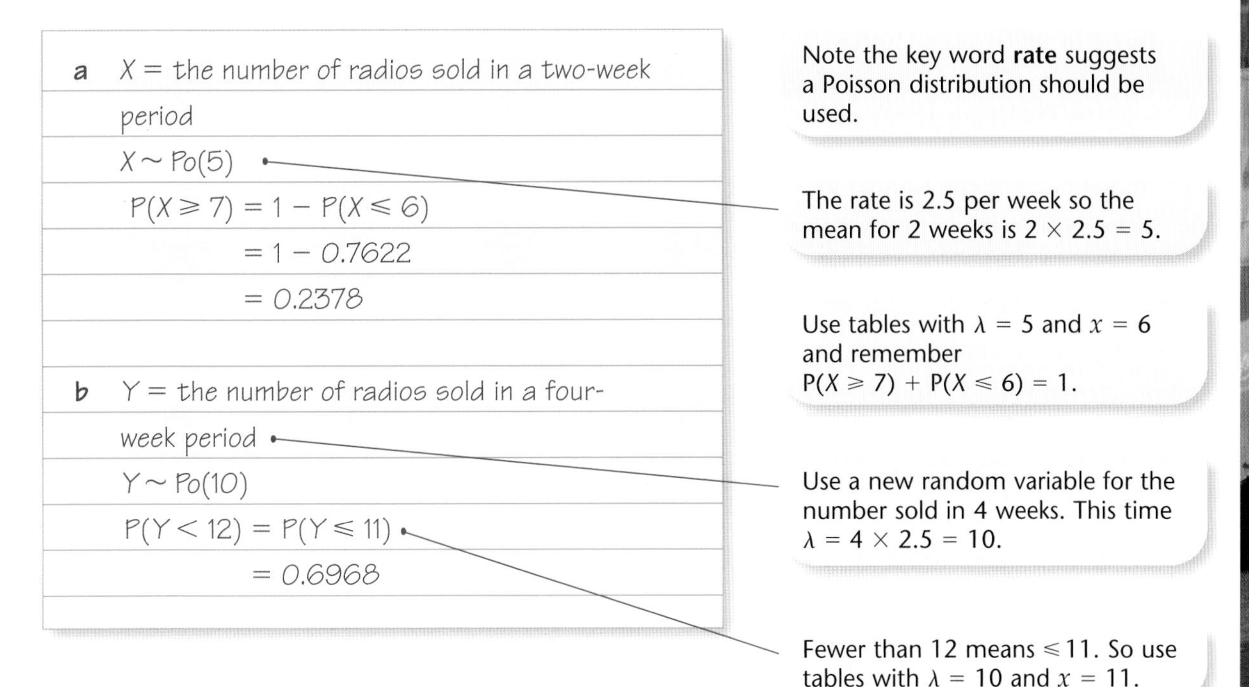

c Let s = the number the manager should have in stock.

The shop will run out of radios if $Y > s$

The manager requires $P(Y > s) < 0.01$

i.e. $1 - P(Y \leqslant s) < 0.01$

or $0.99 < P(Y \leqslant s)$

From tables $P(Y \leqslant 18) = 0.9928$

So the manager should ensure that he has 18 radios in stock after the delivery.

Write the manager's condition as a probability statement using Y.

Use $P(Y > s) + P(Y \leqslant s) = 1$.

Use tables with $\lambda = 10$ and search down the column to find the first value of x where $P(X \leqslant x) > 0.99$.

Always write your conclusion in the context of the original question.

Exercise 2C

1 A technician is responsible for a large number of machines. Minor adjustments have to be made to these machines and these occur at random and at a constant rate of 7 per hour. Find the probability that

a in a particular hour the technician makes 4 or fewer adjustments,

b during a half-hour break no adjustments will be required.

2 A textile firm produces rolls of cloth but slight defects sometimes occur. The average number of defects per square metre is 2.5. Use a Poisson distribution to calculate the probability that

a a 1.5 m^2 portion of cloth bought to make a skirt contains no defects,

b a 4 m^2 portion of cloth contains fewer than 5 defects.

c State briefly what assumptions have to be made before a Poisson distribution can be accepted as a suitable model in this situation.

3 State which of the following could be modelled by a Poisson distribution and which can not. Give reasons for your answers.

a The number of misprints on this page in the first draft of this book.

b The number of pigs in a particular 5 m square of their field 1 hour after their food was placed in a central trough.

c The number of pigs in a particular 5 m square of their field 1 minute after their food was placed in a central trough.

d The amount of salt, in mg, contained in 1 cm^3 of water taken from a bucket immediately after a teaspoon of salt was added to the bucket.

e The number of marathon runners passing the finishing post between 20 and 21 minutes after the winner of the race.

4 The number of accidents per week at a certain road intersection has a Poisson distribution with parameter 2.5. Find the probability that

a exactly 5 accidents will occur in a particular week,

b more than 14 accidents will occur in 4 weeks. *E*

5 In a particular district it has been found, over a long period, that the number, X, of cases of measles reported per month has a Poisson distribution with parameter 1.5. Find the probability that in this district

a in any given month, exactly 2 cases of measles will be reported,

b in a period of 6 months, fewer than 10 cases of measles will be reported.

6 A biologist is studying the behaviour of sheep in a large field. The field is divided into a number of equally sized squares and the average number of sheep per square is 2.5. The sheep are randomly scattered throughout the field.

a Suggest a suitable model for the number of sheep in a square and give a value for any parameter or parameters required.

b Calculate the probability that a randomly selected square contains more than 3 sheep.

A sheep dog has been sent into the field to round up the sheep.

c Explain why the model may no longer be applicable.

7 During office hours, telephone calls to a single telephone in an office come in at an average rate of 18 calls per hour. Assuming that a Poisson distribution can be applied, find the probability that in a 5-minute period there will be

a fewer than 2 calls,

b more than 3 calls.

c Find the probability of no calls during a 20-minute coffee break.

8 A shop sells large birthday cakes at a rate of 2 every 3 days.

a Find the probability of selling no large birthday cakes on a randomly selected day.

Fresh cakes are baked every 3 days and any cakes older than 3 days can not be sold.

b Find how many large birthday cakes should be baked so that the probability of running out of large birthday cakes to sell is less than 1%.

9 On a typical summer's day a boat company hires rowing boats at a rate of 9 per hour.

a Find the probability of hiring out at least 6 boats in a randomly selected 30-minute period.

The company has 8 boats to hire and decides to hire them out for 20-minute periods.

b Show that the probability of running out of boats is less than 1%.

c Find how many boats the company should have to be 99% sure of meeting all demands if the hire period is extended to 30 minutes.

10 Breakdowns on a particular machine occur at random at a rate of 1.5 per week.

a Find the probability that no more than 2 breakdowns occur in a randomly chosen week.

b Find the probability of at least 5 breakdowns in a randomly chosen two-week period.

A maintenance firm offers a contract for repairing breakdowns over a six-week period. The firm will give a full refund if there are more than n breakdowns in a six-week period. The firm want the probability of having to pay a refund to be 5% or less.

c find the smallest value of n.

2.5 In some cases you can approximate a binomial distribution with a Poisson distribution.

Evaluating binomial probabilities when n is large can be quite difficult and in such circumstances it is sometimes simpler to use an approximation.

If $X \sim B(n, p)$ and

- **n is large**
- **p is small**

then X can be approximated by Po(np).

Recall that the mean of a binomial distribution is np. The Poisson approximation has the same mean as the original binomial.

There is no clear rule as to what constitutes 'large n' or 'small p' but usually the value for np will be $\leqslant 10$ so that the Poisson table can be used. Generally the larger the value of n and the smaller the value of p the better the approximation will be.

If the original binomial distribution can be used then it is always best to do so *unless* the question specifically instructs you to use an approximation.

Example 7

The probability that my favourite make of chocolate biscuit is 'double wrapped' is 0.01. Use a suitable approximation to find the probability that in a box of 60 biscuits

a none are double wrapped,

b at least 2 are double wrapped.

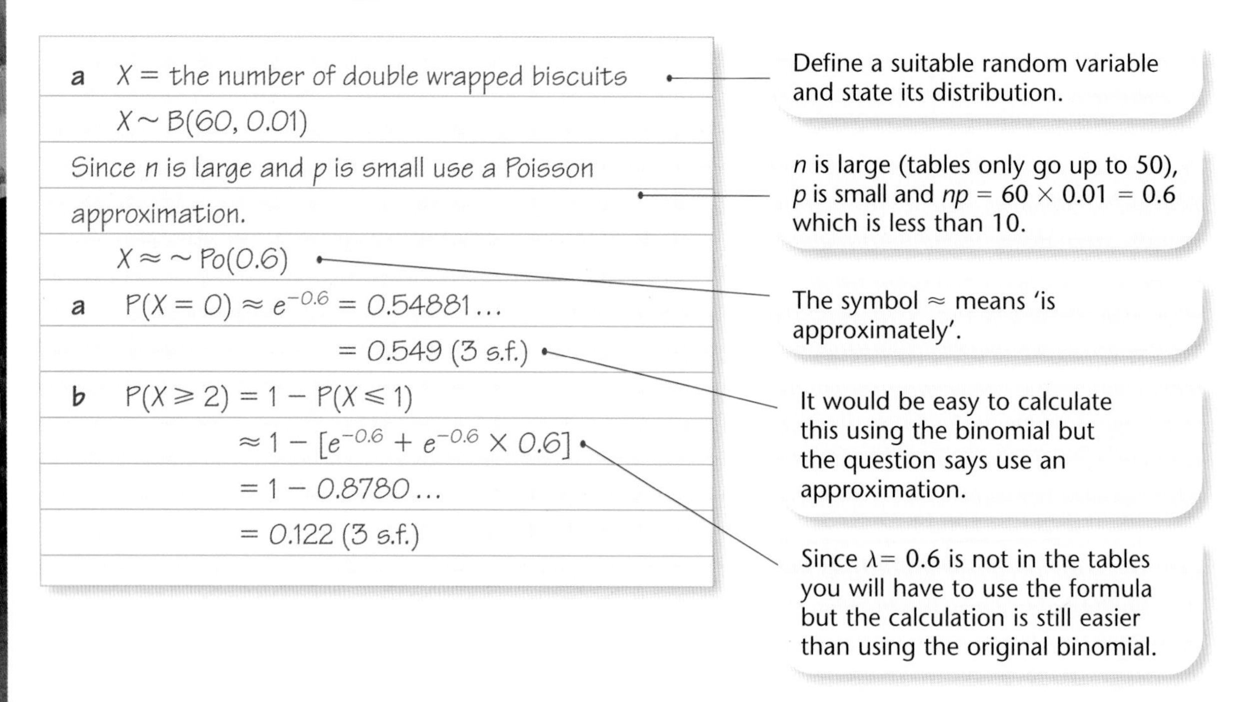

a X = the number of double wrapped biscuits

$X \sim B(60, 0.01)$

Since n is large and p is small use a Poisson approximation.

$X \approx\sim \text{Po}(0.6)$

a $P(X = 0) \approx e^{-0.6} = 0.54881\ldots$

$= 0.549$ (3 s.f.)

b $P(X \geqslant 2) = 1 - P(X \leqslant 1)$

$\approx 1 - [e^{-0.6} + e^{-0.6} \times 0.6]$

$= 1 - 0.8780\ldots$

$= 0.122$ (3 s.f.)

Define a suitable random variable and state its distribution.

n is large (tables only go up to 50), p is small and $np = 60 \times 0.01 = 0.6$ which is less than 10.

The symbol $\approx$ means 'is approximately'.

It would be easy to calculate this using the binomial but the question says use an approximation.

Since $\lambda = 0.6$ is not in the tables you will have to use the formula but the calculation is still easier than using the original binomial.

NB The answers using the original binomial distribution are **a** 0.547 and **b** 0.121, so you can see that the approximation works quite accurately.

It is important to remember that ***p* should be small before the approximation is used**. One reason for this can be seen by considering the variance of the random variable.

If $X \sim B(n, p)$ then $\mu = np$ and $\sigma^2 = np(1 - p)$. If p is small then $(1 - p)$ will be close to 1 and so $\sigma^2 \approx np = \lambda$ and you saw (in Section 2.2) that for a Poisson variable $E(X) = Var(X)$. However an approximation can be used if p is close to 1 by using the complementary random variable (counting the number of failures instead of the number of successes).

Example 8

A garden centre states that 95% of its daffodil bulbs will produce flowers the following season. Gill buys 100 bulbs. Use a suitable approximation to estimate the probability that at least 96 of the bulbs will flower next season.

Let X = the number of bulbs that flower

$X \sim B(100, 0.95)$

Let Y = the number of bulbs that do not flower

$Y \sim B(100, 0.05)$

Using a Poisson approximation

$Y \approx \sim Po(5)$

$P(X \geqslant 96) = P(Y \leqslant 4)$

$= 0.4405$

Define a suitable random variable.

n is large but p is close to 1 so a Poisson approximation cannot be used.

Define the complementary variable.

Now n is large, p is small and $np = 5 < 10$ so a Poisson approximation can be used.

At least 96 flower means $X \geqslant 96$ and this means that 4 or fewer don't so $Y \leqslant 4$.

Use tables with $\lambda = 5$ and $x = 4$.

Some calculators can find the binomial probability (0.43598…) very easily without the need to use an approximation or the complementary variable. This can provide a useful check in the examination but if the question tells you to use a suitable approximation you will gain no credit for using a calculator to find the binomial probability.

Exercise 2D

1 The random variable $X \sim B(80, 0.10)$. Using a suitable approximation, find

a $P(X \geqslant 1)$, **b** $P(X \leqslant 6)$.

2 The random variable $X \sim B(120, 0.02)$. Using a suitable approximation, find

a $P(X = 1)$, **b** $P(X \geqslant 3)$.

3 The random variable $X \sim B(50, 0.05)$. Find the percentage error in $P(X \leqslant 4)$ when X is approximated by a Poisson distribution.

4 In a certain manufacturing process the proportion of defective articles produced is 2%. In a batch of 300 articles, use a suitable approximation to find the probability that

a there are fewer than 2 defectives,

b there are exactly 4 defectives.

5 A medical practice screens a random sample of 250 of its patients for a certain condition which is present in 1.5% of the population. Use a suitable approximation to find the probability that they obtain

a no patients with the condition,

b at least two patients with the condition.

6 An experiment involving 2 fair dice is carried out 180 times. The dice are placed in a container, shaken and the number of times a double six is obtained recorded. Use a suitable approximation to find the probability that a double six is obtained

a once, **b** twice, **c** at least three times.

7 It is claimed that 95% of the population in a certain village are right-handed. A random sample of 80 villagers is tested to see whether or not they are right-handed. Use a Poisson approximation to estimate the probability that the number who are right-handed is

a 80, **b** 79, **c** at least 78.

8 In a computer simulation 500 dots were fired at a target and the probability of a dot hitting the target was 0.98. Find the probability that

a all the dots hit the target, **b** at least 495 hit the target.

9 **a** State the conditions under which the Poisson distribution may be used as an approximation to the binomial distribution.

Independently for each call into the telephone exchange of a large organisation, there is a probability of 0.002 that the call will be connected to a wrong extension.

b Find, to 3 significant figures, the probability that, on a given day, exactly one of the first 5 incoming calls will be wrongly connected.

c Use a Poisson approximation to find, to 3 decimal places, the probability that, on a day when there are 1000 incoming calls, at least 3 of them are wrongly connected during that day. *E*

2.6 Deciding whether a binomial or a Poisson distribution is an appropriate model.

- There are certain **key words** to look for:

  mention of **rate** ... suggests a Poisson distribution
  mention of a **fixed number** of trials or a **proportion** suggests a binomial distribution.

- There are certain **key questions** to ask.

  Is there a value for *n* – the number of trials? If the answer is 'yes', then use a binomial distribution.

  Is there a constant probability of success or proportion? If the answer is 'yes', then use a binomial distribution

  Is there an average number of occurrences or a rate of occurrences? If the answer is 'yes', then use a Poisson distribution.

  Can you say how many times something will *not* happen? If you can, then there is probably a fixed *n* and a binomial distribution can be used, if not then it is probably a Poisson distribution.

Example 9

A bakery claims that a pack of 10 of their teacakes contains on average 75 currants.

a Suggest a distribution that could be used to model the number of currants in a randomly selected teacake and state the value of any parameters.

State any assumptions that must be made for the model to be valid.

b Find the probability that a randomly selected teacake contains more than 7 currants.

a Let X = the number of currants in a randomly selected teacake.

$X \sim \text{Po}(7.5)$

There is no fixed number of currants but an *average* number per teacake of 7.5.

You must ensure that currants are randomly distributed. (The mixture is well stirred.)

Refer back to Section 2.4 and write your answer in context.

b $P(X > 7) = 1 - P(X \leq 7)$

$= 1 - 0.5246$

$= 0.4754$

Rewrite $P(X > 7)$ as $1 - P(X \leq x)$ and use tables with $\lambda = 7.5$ and $x = 7$.

Sometimes a question can involve both binomial and Poisson distributions because the random variable being considered changes.

Example 10

A piece of machinery breaks down at a rate of once a fortnight, but these breakdowns occur randomly.

a Find the probability of a week with no breakdowns, giving your answer to 3 d.p.

A week with no breakdowns is called a star week. Every 12 weeks the number of star weeks is recorded and a report is sent to the machine manufacturers.

b Find the probability that there are more than 10 star weeks in the next report. Give your answer to 2 d.p.

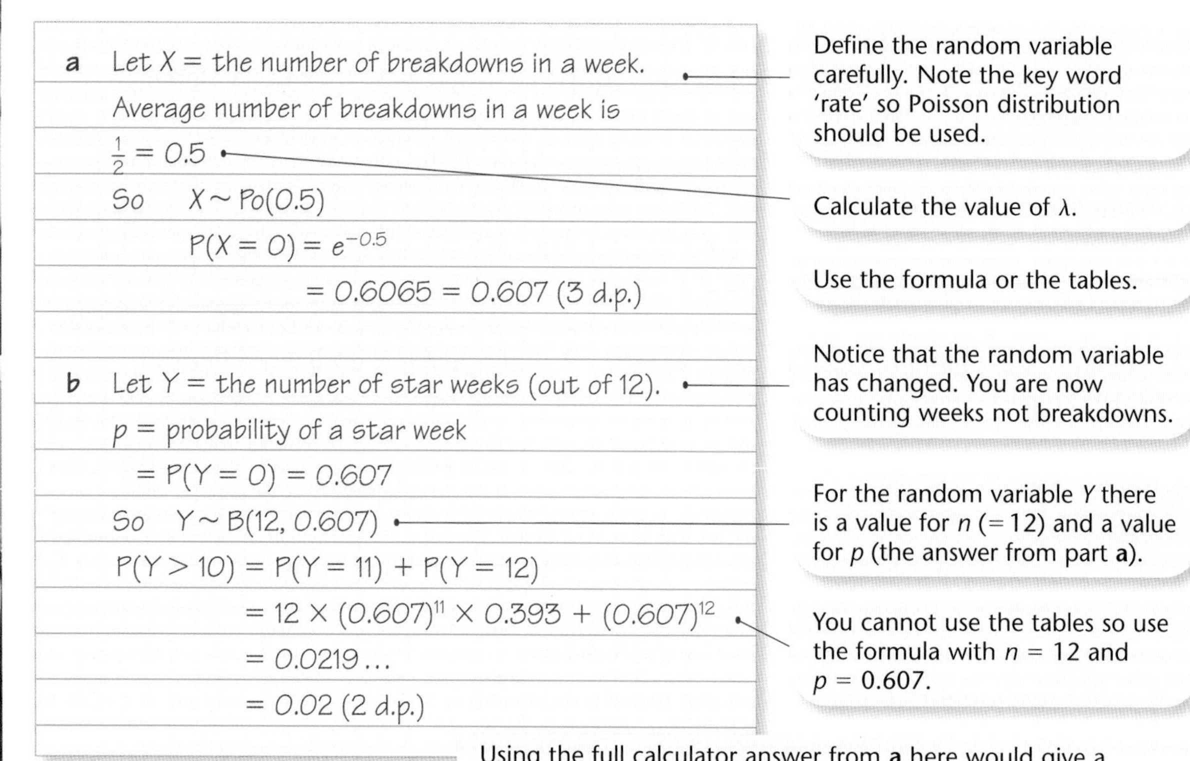

a Let X = the number of breakdowns in a week.

Average number of breakdowns in a week is $\frac{1}{2} = 0.5$

So $X \sim \text{Po}(0.5)$

$P(X = 0) = e^{-0.5}$

$= 0.6065 = 0.607$ (3 d.p.)

Define the random variable carefully. Note the key word 'rate' so Poisson distribution should be used.

Calculate the value of λ.

Use the formula or the tables.

b Let Y = the number of star weeks (out of 12).

p = probability of a star week

$= P(Y = 0) = 0.607$

So $Y \sim B(12, 0.607)$

$P(Y > 10) = P(Y = 11) + P(Y = 12)$

$= 12 \times (0.607)^{11} \times 0.393 + (0.607)^{12}$

$= 0.0219\ldots$

$= 0.02$ (2 d.p.)

Notice that the random variable has changed. You are now counting weeks not breakdowns.

For the random variable Y there is a value for n (= 12) and a value for p (the answer from part **a**).

You cannot use the tables so use the formula with $n = 12$ and $p = 0.607$.

Using the full calculator answer from **a** here would give a slightly more accurate final answer to **b** but it would still be 0.02 to 2 d.p. In an examination, if an answer is given in part **a** it is acceptable to use it to that level of accuracy in part **b**.

Exercise 2E

1 **a** State conditions under which the Poisson distribution is a suitable model to use in statistical work.

Flaws in a certain brand of tape occur at random and at a rate of 0.75 per 100 metres. Assuming a Poisson distribution for the number of flaws in a 400 metre roll of tape,

b find the probability that there will be at least one flaw.

c Show that the probability that there will be at most 2 flaws is 0.423 (to 3 decimal places).

In a batch of 5 rolls, each of length 400 metres,

d find the probability that at least 2 rolls will contain fewer than 3 flaws.

2 An archer fires arrows at a target and for each arrow, independently of all others, the probability that it hits the bull's eye is $\frac{1}{8}$.

a Given that the archer fires 5 arrows, find the probability that fewer than 2 arrows hit the bull's eye.

The archer fires 5 arrows, collects them and then fires all 5 again.

b Find the probability that on both occasions fewer than 2 hit the bull's eye.

The archer now fires 60 arrows at the target. Using a suitable approximation find

c the probability that fewer than 10 hit the bull's eye,

d the greatest value of m such that the probability that the archer hits the bull's eye with at least m arrows is greater than 0.5.

3 In Joe's roadside café $\frac{2}{5}$ of the customers buy a cup of tea.

a Find the probability that at least 4 of the next 10 customers will buy a cup of tea.

Joe has calculated that, on a typical morning, customers arrive in the café at a rate of 0.5 per minute.

b Find the probability that at least 10 customers arrive in the next 15 minutes.

c Find the probability that exactly 10 customers arrive in the next 20 minutes.

d Find the probability that in the next 20 minutes exactly 10 customers arrive and at least 4 of them buy a cup of tea.

4 The number, X, of breakdowns per week of the lifts in a large block of flats has a Poisson distribution with mean 0.25. Find, to 3 decimal places, the probability that in a particular week

a there will be at least one breakdown,

b there will be at most 2 breakdowns.

c Show that the probability that during a 12-week period there will be no lift breakdowns is 0.050 (to 3 decimal places).

The residents in the flats have a maintenance contract with *Liftserve*. The contract is for a set of 20, 12-week periods. For every 12-week period with no breakdowns the residents pay *Liftserve* £500. If there is at least 1 breakdown in a 12-week period then *Liftserve* will mend the lift free of charge and the residents pay nothing for that period of 12 weeks.

d Find the probability that over the course of the contract the residents pay no more than £1000.

5 Accidents occur in a school playground at the rate of 3 per year.

a Suggest a suitable model for the number of accidents in the playground next month.

b Using this model calculate the probability of 1 or more accidents in the playground next month.

Mixed exercise 2F

1 During working hours an office switchboard receives telephone calls at random and at a rate of one call every 40 seconds.

a Find, to 3 decimal places, the probability that during a given one-minute period

i no call is received, **ii** at least 2 calls are received.

b Find, to 3 decimal places, the probability that no call is received between 10:30 a.m. and 10:31 a.m. and that at least two calls are received between 10:31 a.m. and 10:32 a.m. **E**

2 State conditions under which the Poisson distribution is a suitable model to use in statistical work.

The number of typing errors per 1000 words made by a typist has a Poisson distribution with mean 2.5.

a Find, to 3 decimal places, the probability that in an essay of 4000 words there will be at least 12 typing errors.

The typist types 3 essays, each of length 4000 words.

b Find the probability that each contains at least 12 typing errors. **E**

3 **a** State conditions under which the binomial distribution $B(n, p)$ may be approximated by a Poisson distribution and write down the mean of this Poisson distribution.

Samples of blood were taken from 250 children in a region of India. Of these children, 4 had blood type $A2B$.

b Write down an estimate of p, the proportion of children in this region having blood type $A2B$.

Consider a group of n children from this region and let X be the number having blood type $A2B$. Assuming that X is distributed $B(n, p)$ and that p has the value estimated above, calculate, to 3 decimal places, the probability that the number of children with blood type $A2B$ in a group of 6 children from this region will be

i zero, **ii** more than 1.

c Use a Poisson approximation to calculate, to 4 decimal places, the probability that, in a group of 800 children from this region, there will be fewer than 3 children of blood type $A2B$. *E*

4 Which of the following variables is best modelled by a Poisson distribution and which is best modelled by a binomial distribution?

a The number of hits by an arrow on a target, when 20 arrows are fired.

b The number of earth tremors that take place in a village over a given period of time.

c The number of particles emitted per minute by a radioactive isotope.

d The number of heads you get when tossing 2 coins 100 times.

e The number of accidents in a city in a year.

f The number of flying bomb hits in specified areas of London during World War 2.

5 Loaves of bread on a production line pass a monitoring point at a constant rate of 300 loaves per hour.

a Find how many loaves you would expect to pass the monitoring point in 2 minutes.

b Find the probability that no loaves pass the monitoring point in a given 1-minute period.

6 Accidents occur at a certain road junction at a rate of 3 per year.

a Suggest a suitable model for the number of accidents at this road junction in the next month.

b Show that, under this model, the probability of 2 or more accidents at this road junction in the next month is 0.0265 to 4 decimal places.

The local residents have applied for a crossing to be installed.

The planning committee agree to monitor the situation for the next 12 months. If there is at least one month with 2 or more accidents in it they will install a crossing.

c Find the probability that the crossing is installed. *E*

7 Breakdowns occur on a particular machine at a rate of 2.5 per month. Assuming that the number of breakdowns can be modelled by a Poisson distribution, find the probability that

a exactly 3 occur in a particular month,

b more than 10 occur in a three-month period,

c exactly 3 occur in each of 2 successive months. *E*

8 A geography student is studying the distribution of telephone boxes in a large rural area where there is an average of 300 boxes per 500 km^2. A map of part of the area is divided into 50 squares, each of area 1 km^2 and the student wishes to model the number of telephone boxes per square.

a Suggest a suitable model the student could use and specify any parameters required.

One of the squares is picked at random.

b Find the probability that this square does not contain any telephone boxes.

c Find the probability that this square contains at least 3 telephone boxes.

The student suggests using this model on another map of a large city and surrounding villages.

d Comment, giving your reason briefly, on the suitability of the model in this situation. *E*

9 All the letters in a particular office are typed either by Pat, a trainee typist, or by Lyn, who is a fully-trained typist. The probability that a letter typed by Pat will contain one or more errors is 0.3.

a Find the probability that a random sample of 4 letters typed by Pat will include exactly one letter free from error.

The probability that a letter typed by Lyn will contain one or more errors is 0.05.

b Use tables, or otherwise, to find, to 3 decimal places, the probability that in a random sample of 20 letters typed by Lyn, not more than 2 letters will contain one or more errors.

On any one day, 6% of the letters typed in the office are typed by Pat. One letter is chosen at random from those typed on that day.

c Show that the probability that it will contain one or more errors is 0.065.

Given that each of 2 letters chosen at random from the day's typing contains one or more errors,

d find, to 4 decimal places, the probability that one was typed by Pat and the other by Lyn. *E*

10 The number of breakdowns per day of the lifts in a large block of flats is modelled by a Poisson distribution with mean 0.2.

a Find, to 3 decimal places, the probability that on a particular day there will be at least one breakdown.

b Find the probability that there are fewer than 2 days in a 30-day month with at least one breakdown.

Summary of key points

1 If $X \sim \text{Po}(\lambda)$ then $\text{P}(X = x) = \text{e}^{-\lambda}\frac{\lambda^x}{x!}$

2 If $X \sim \text{Po}(\lambda)$ then $\text{E}(X) = \lambda$ and $\text{Var}(X) = \lambda$

3 If $X \sim \text{B}(n, p)$ and n is large and p is small then $X \approx\sim \text{Po}(\lambda)$, where $\lambda = np$

After completing this chapter you should

- be able to decide whether something can be a probability density function
- be able to find the cumulative distribution function given the probability density function
- be able to find the probability density function given the cumulative distribution function
- be able to find the mean and variance of a random variable using its probability density function
- be able to find the mode and median of a random variable using its probability density function.

3 Continuous random variables

The duration, t, of phone calls, in minutes, is modelled by the function

$$f(t) = \begin{cases} \frac{1}{k}(t-2), & 2 \leqslant t \leqslant 12 \\ 0, & \text{otherwise} \end{cases}$$

Bill makes a call. What is the probability the call lasts between 3 and 5 minutes?
You will be able to answer questions such as this when you have studied this chapter.

3.1 The concept of a continuous random variable and its probability density function.

In book S1, Chapter 4, you looked at continuous random variables.
For example, the histogram below shows the variable t, the time in seconds taken by 100 children to do their homework.

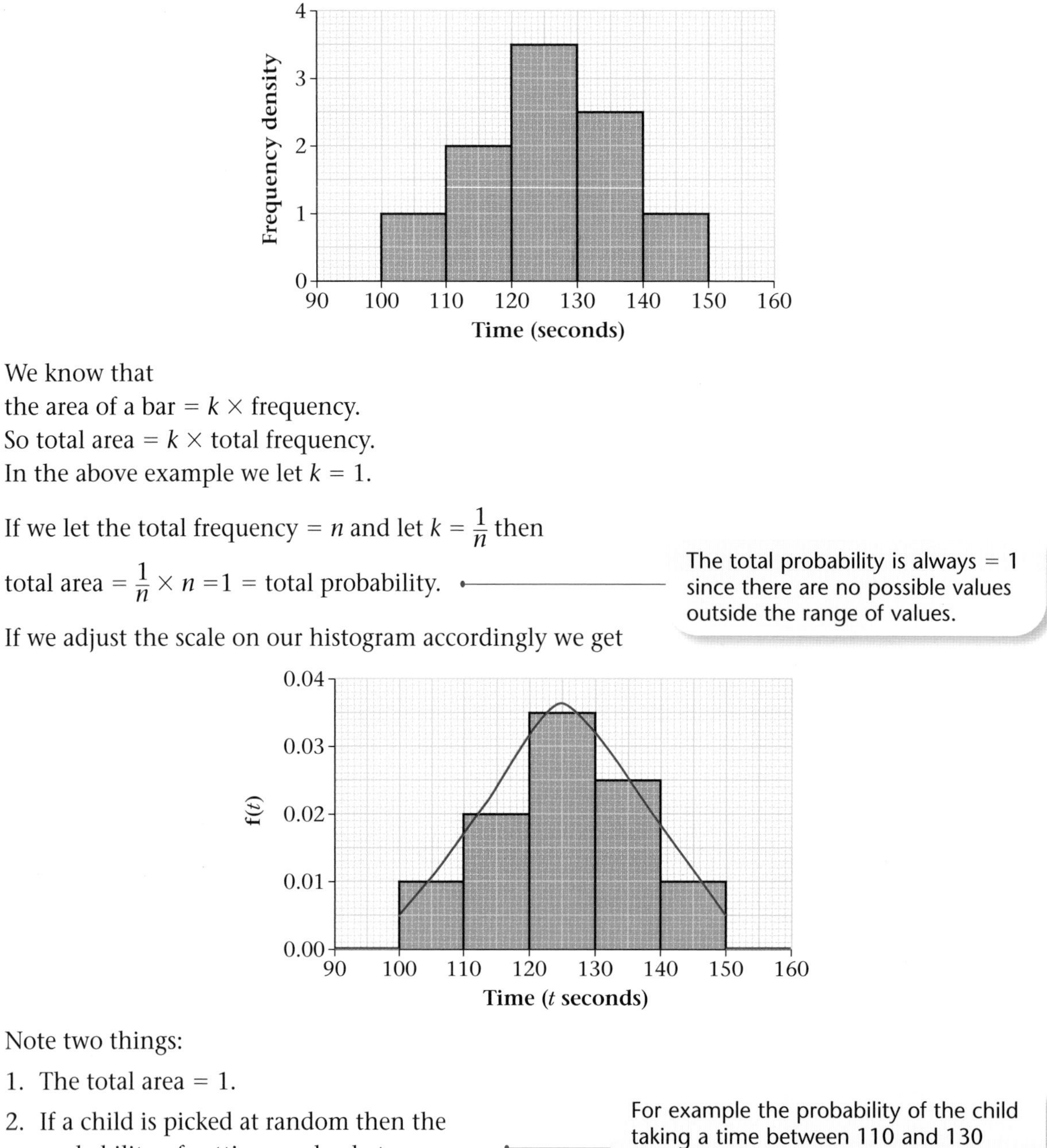

We know that
the area of a bar = $k \times$ frequency.
So total area = $k \times$ total frequency.
In the above example we let $k = 1$.

If we let the total frequency = n and let $k = \frac{1}{n}$ then

total area = $\frac{1}{n} \times n = 1$ = total probability.

The total probability is always = 1 since there are no possible values outside the range of values.

If we adjust the scale on our histogram accordingly we get

Note two things:

1. The total area = 1.
2. If a child is picked at random then the probability of getting a value between any two times is now the area between those two times.

For example the probability of the child taking a time between 110 and 130 seconds is
area = $10 \times 0.02 + 10 \times 0.035 = 0.55$.

A probability density function f(x) is a smooth version of a histogram that is drawn in the above way. (See red line.)

A probability density function, p.d.f. for short, has a few important properties.

- **If X is a continuous random variable with p.d.f. $f(x)$ then**

  1. $f(x) \geqslant 0$ since we cannot have negative probabilities.

  2. $P(a < X < b)$ = shaded area

  $$= \int_a^b f(x)\,dx$$

  3. $\int_{-\infty}^{\infty} f(x)\,dx = 1$ since the area under the curve = 1

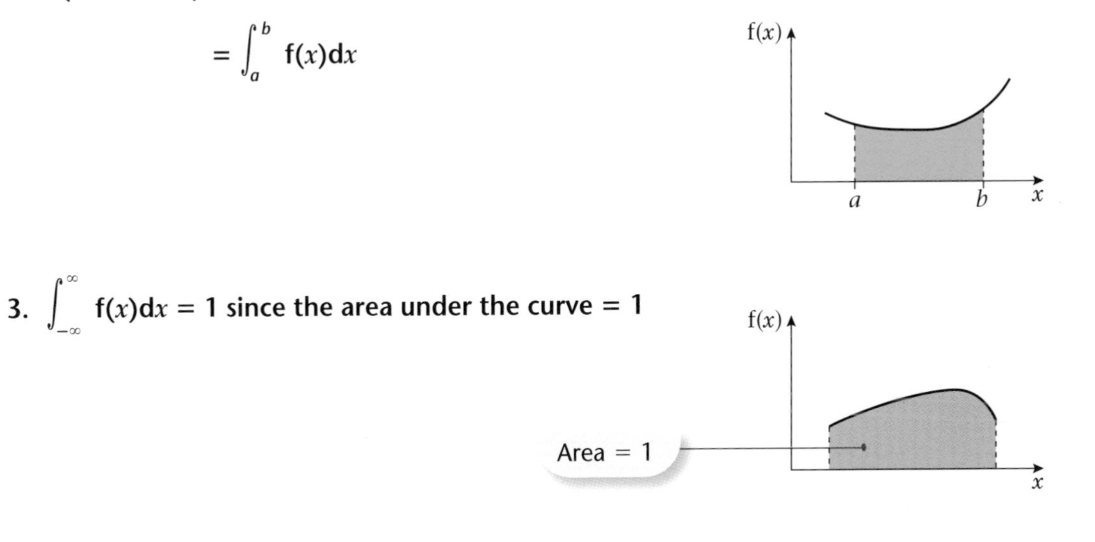

Example 1

Which of the following could be a probability density function? Give a reason for your answer.

a $f(x) = \begin{cases} 2x, & -2 \leqslant x \leqslant 3, \\ 0, & \text{otherwise.} \end{cases}$

b $f(x) = \begin{cases} k(x-2), & 3 \leqslant x \leqslant 5, \\ 0, & \text{otherwise.} \end{cases}$

c $f(x) = \begin{cases} kx(x-2), & 1 \leqslant x \leqslant 3, \\ 0, & \text{otherwise.} \end{cases}$

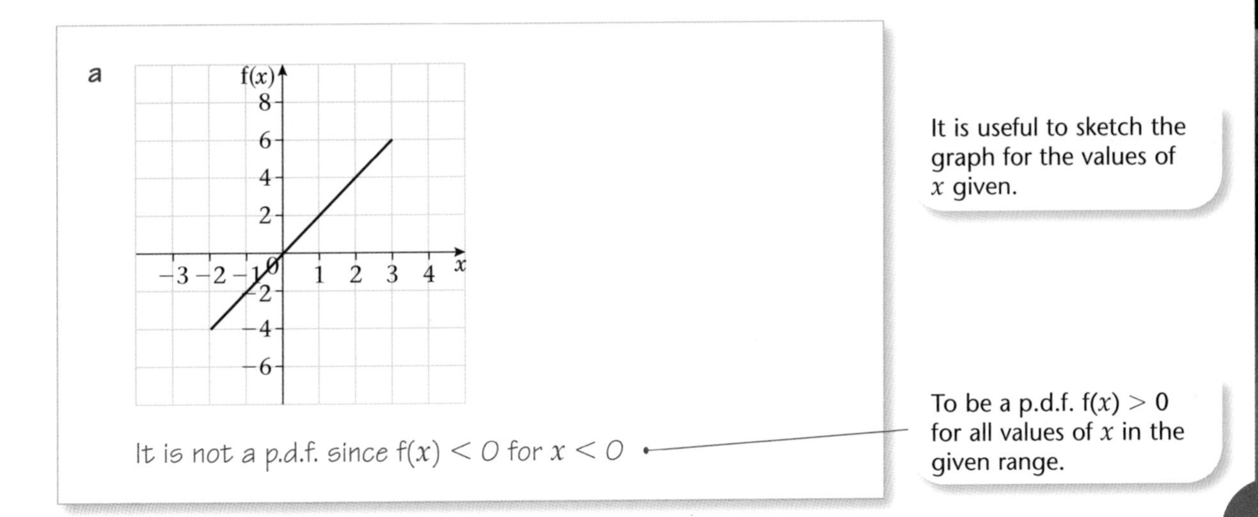

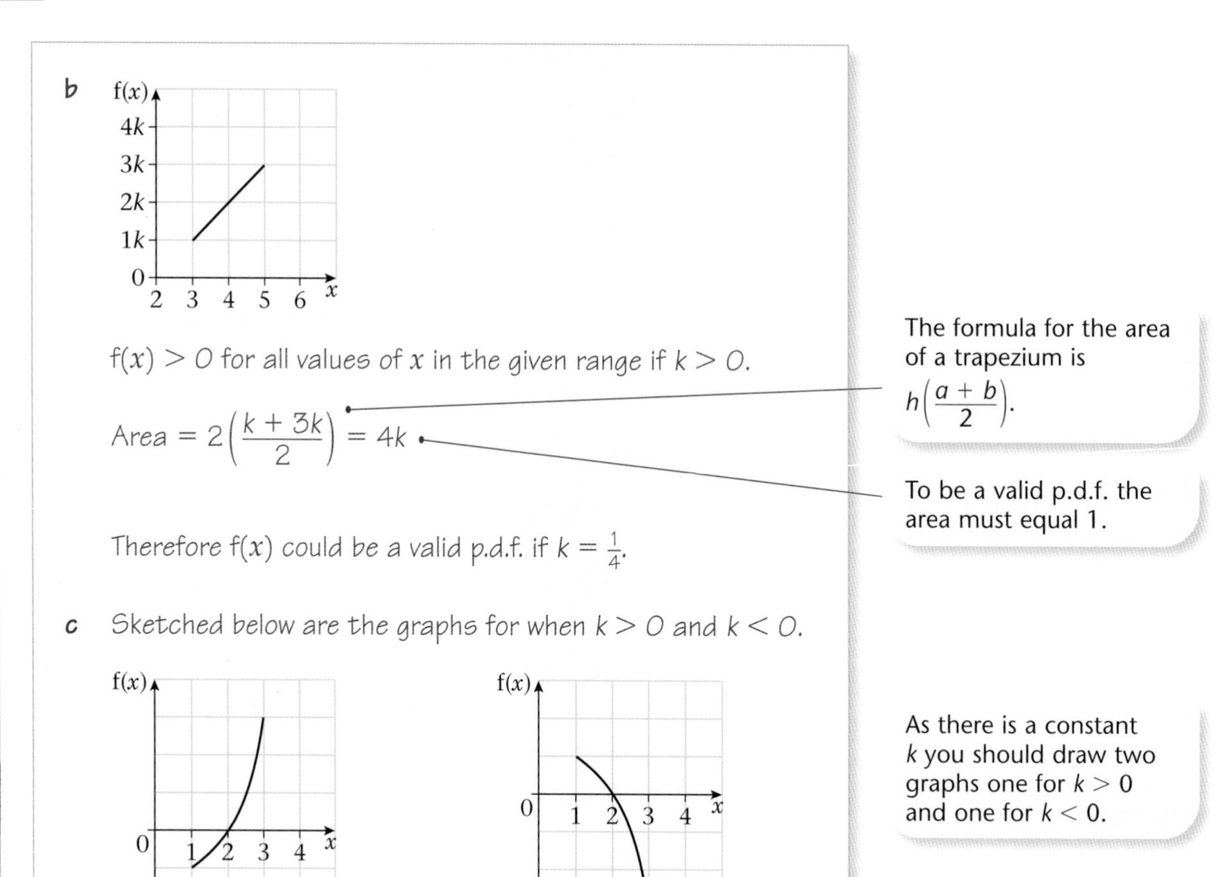

b $f(x) > 0$ for all values of x in the given range if $k > 0$.

$$\text{Area} = 2\left(\frac{k + 3k}{2}\right) = 4k$$

The formula for the area of a trapezium is $h\left(\frac{a + b}{2}\right)$.

To be a valid p.d.f. the area must equal 1.

Therefore $f(x)$ could be a valid p.d.f. if $k = \frac{1}{4}$.

c Sketched below are the graphs for when $k > 0$ and $k < 0$.

graph if $k > 0$ graph if $k < 0$

As there is a constant k you should draw two graphs one for $k > 0$ and one for $k < 0$.

So for any value of k there is some value of x in the given range such that $f(x) < 0$.

Therefore $f(x)$ cannot be a probability density function.

Example 2

The random variable X has probability density function

$$f(x) = \begin{cases} kx(4 - x), & 2 \leqslant x \leqslant 4, \\ 0, & \text{otherwise.} \end{cases}$$

Find the value of k and sketch the p.d.f.

$$\int_2^4 k(4x - x^2)dx = 1$$

Area under the curve must equal 1.

$$\int_2^4 f(x)dx = 1$$

$$k\left[2x^2 - \frac{x^3}{3}\right]_2^4 = 1$$

Remember $\int x^n dx = \frac{x^{n+1}}{n+1}$.

$$k\left[\left(32 - \frac{64}{3}\right) - \left(8 - \frac{8}{3}\right)\right] = 1$$

When substituting the limits it is the value with the top limit substituted minus the value with the bottom limit substituted.

$$k\left(\frac{16}{3}\right) = 1$$

$$k = \left(\frac{3}{16}\right)$$

Sketching the graph gives

When sketching the graph remember to label the axes and label the important values. These are the boundaries of the given range of x values and their corresponding y values. Here they are 2, 4 and 0.75.

You must make it clear when $f(x) = 0$.

Example 3

The random variable X has probability density function

$$f(x) = \begin{cases} k, & 1 < x < 2, \\ k(x-1), & 2 \leqslant x \leqslant 4, \\ 0, & \text{otherwise.} \end{cases}$$

a Sketch $f(x)$.

b Find the value of k.

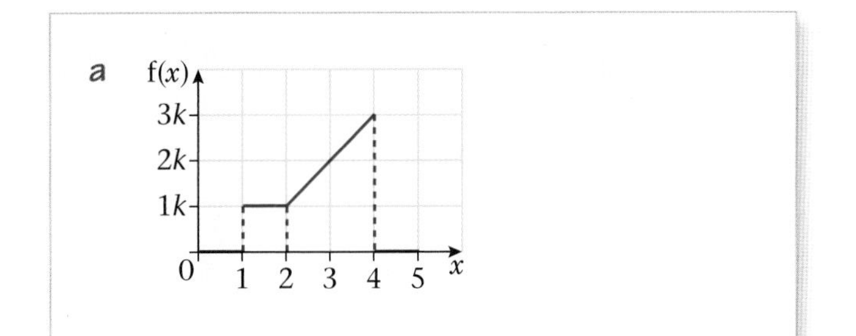

b $$\int_1^2 k\,dx + \int_2^4 k(x-1)\,dx = 1$$

$$[kx]_1^2 + \left[\frac{kx^2}{2} - kx\right]_2^4 = 1$$

$$k + [(8k - 4k) - (2k - 2k)] = 1$$

$$5k = 1$$

$$k = \tfrac{1}{5}$$

The total area = 1. Here the area has been found by integrating but it is sometimes easier to find the value of k by looking at the sketch.

Area = $k + \frac{1}{2} \times 2 \times (k + 3k) = 5k$

(using the area of a rectangle and trapezium)

Area = 1

$5k = 1$

$k = \frac{1}{5}$

Exercise 3A

1 Give reasons why the following are not valid probability density functions.

a $f(x) = \begin{cases} \frac{1}{4}x, & -1 \leqslant x \leqslant 2, \\ 0, & \text{otherwise.} \end{cases}$

b $f(x) = \begin{cases} x^2, & 1 \leqslant x \leqslant 3, \\ 0, & \text{otherwise.} \end{cases}$

c $f(x) = \begin{cases} (x^3 - 2), & -1 \leqslant x \leqslant 3, \\ 0, & \text{otherwise.} \end{cases}$

2 For what value of k is the following a valid probability density function?

$f(x) = \begin{cases} k(1 - x^2), & -4 \leqslant x \leqslant -2, \\ 0, & \text{otherwise.} \end{cases}$

3 Sketch the following probability density functions.

a $f(x) = \begin{cases} \frac{1}{8}(x - 2), & 2 \leqslant x \leqslant 6, \\ 0, & \text{otherwise.} \end{cases}$

b $f(x) = \begin{cases} \frac{2}{15}(5 - x), & 1 \leqslant x \leqslant 4, \\ 0, & \text{otherwise.} \end{cases}$

4 Find the value of k so that each of the following are valid probability density functions.

a $f(x) = \begin{cases} kx, & 1 \leqslant x \leqslant 3, \\ 0, & \text{otherwise.} \end{cases}$

b $f(x) = \begin{cases} kx^2, & 0 \leqslant x \leqslant 3, \\ 0, & \text{otherwise.} \end{cases}$

c $f(x) = \begin{cases} k(1 + x^2), & -1 \leqslant x \leqslant 2, \\ 0, & \text{otherwise.} \end{cases}$

5 The continuous random variable X has probability density function given by:

$$f(x) = \begin{cases} k(4 - x), & 0 \leqslant x \leqslant 2, \\ 0, & \text{otherwise.} \end{cases}$$

a Find the value of k.

b Sketch the probability density function for all values of x.

6 The continuous random variable X has probability density function given by:

$$f(x) = \begin{cases} kx^2(2 - x), & 0 \leqslant x \leqslant 2, \\ 0, & \text{otherwise.} \end{cases}$$

Find the value of k.

7 The continuous random variable X has probability density function given by:

$$f(x) = \begin{cases} kx^3, & 1 \leqslant x \leqslant 4, \\ 0, & \text{otherwise.} \end{cases}$$

Find the value of k.

8 The continuous random variable X has probability density function given by:

$$f(x) = \begin{cases} k, & 0 < x < 2, \\ k(2x - 3), & 2 \leqslant x \leqslant 3, \\ 0, & \text{otherwise.} \end{cases}$$

a Find the value of k.

b Sketch the probability density function for all values of x.

3.2 The cumulative distribution function.

In book S1, Chapter 8 Section 8.5, you met the cumulative distribution function, $F(x) = P(X \leqslant x)$, for a discrete random variable. If we introduce a continuous random variable T such that $f(t) = f(x)$ then for the continuous random variable X,

$$F(x) = P(X \leqslant x) = \int_{-\infty}^{x} f(t)dt$$

Note: $f(t)$ has the same distribution as $f(x)$ and we use t rather than x to avoid confusion with the limit of integration.

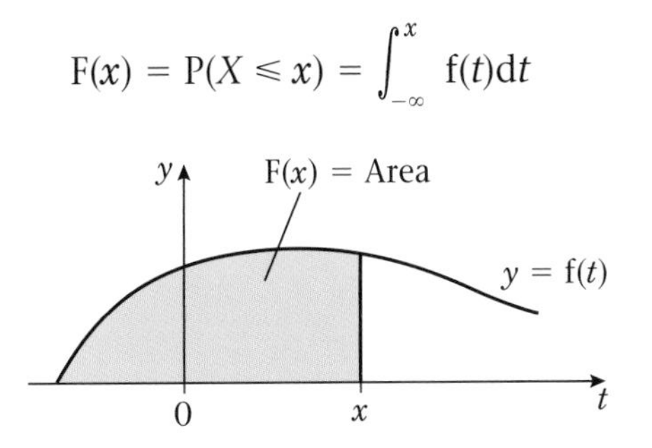

Notice we use the notation $F(x)$ (i.e. capital f) for the cumulative distribution function (c.d.f. for short) but $f(x)$ for the probability density function (p.d.f. for short).

■ **If X is a continuous random variable with c.d.f. $F(x)$ and p.d.f. $f(x)$:**

$$f(x) = \frac{d}{dx}F(x) \text{ and } F(x) = \int_{-\infty}^{x} f(t)dt.$$

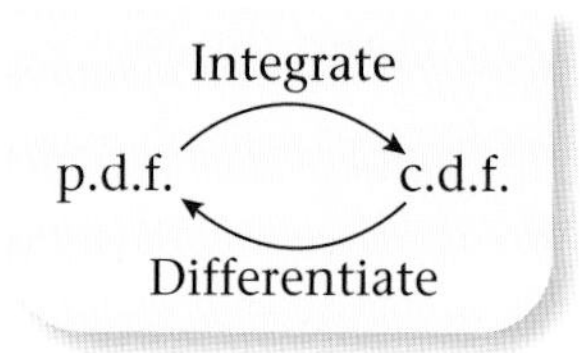

Example 4

The random variable X has probability density function

$$f(x) = \begin{cases} \frac{1}{4}x, & 1 \leqslant x \leqslant 3, \\ 0, & \text{otherwise.} \end{cases}$$

Find $F(x)$.

Method 1

$$F(x) = \int_1^x \frac{1}{4}t\,dt$$

Since between $-\infty$ and 1 $f(x)$ is zero, we use the lower bound of the given interval i.e. one.

$$= \left[\frac{t^2}{8}\right]_1^x$$

$$= \frac{x^2}{8} - \frac{1}{8}$$

Method 2

$$F(x) = \int \frac{1}{4}x\,dx$$

An alternative method is to use an indefinite integral and put + C. We can use $f(x)$ here since there are no limits in the integration.

$$= \frac{x^2}{8} + C$$

$$\frac{3^2}{8} + C = 1$$

C can be found by using $F(3) = 1$. Three is the upper value of the given range,
or
$F(1) = 0$. One is the lower value of the given range.

$$C = -\frac{1}{8}$$

$$F(x) = \begin{cases} 0, & x < 1, \\ \frac{x^2}{8} - \frac{1}{8}, & 1 \leqslant x \leqslant 3, \\ 1, & x > 3. \end{cases}$$

You must define $F(x)$ over the whole range $(-\infty, \infty)$. $F(x) = 0$ for all values less than 1 and $F(x) = 1$ for all values greater than 3 (see diagram).

Example 5

The random variable X has probability density function

$$f(x) = \begin{cases} \frac{1}{5}, & 1 < x < 2, \\ \frac{1}{5}(x-1), & 2 \leqslant x \leqslant 4, \\ 0, & \text{otherwise.} \end{cases}$$

Find F(x).

<u>Method 1</u>

If $x \leqslant 1$

$F(x) = 0 \qquad \text{so } F(1) = 0$

F(x) is cumulative. i.e. F(n) is the P($X \leqslant n$)

If $1 < x < 2$

$$F(x) = F(1) + \int_1^x \frac{1}{5}dt$$

$$= \left[\frac{1}{5}t\right]_1^x$$

$$= \frac{1}{5}x - \frac{1}{5}$$

so $F(2) = \frac{1}{5}$

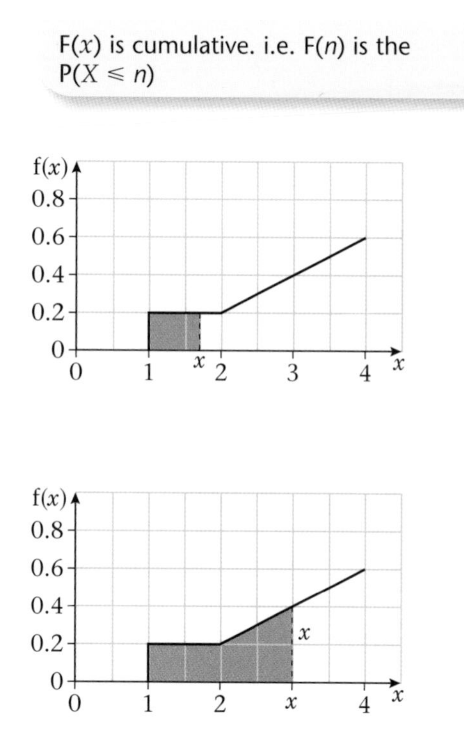

If $2 \leqslant x \leqslant 4$

$$F(x) = F(2) + \int_2^x \frac{1}{5}(t-1)dt$$

$$\frac{1}{5} + \left[\frac{t^2}{10} - \frac{t}{5}\right]_2^x$$

$$= \left[\frac{1}{5}\right] + \left[\left(\frac{x^2}{10} - \frac{x}{5}\right) - \left(\frac{4}{10} - \frac{2}{5}\right)\right]$$

$$= \frac{x^2}{10} - \frac{x}{5} + \frac{1}{5}$$

<u>Method 2</u>

If $1 < x < 2$

$$F(x) = \int \frac{1}{5}dx$$

$$= \frac{1}{5}x + C$$

$$\frac{1}{5} + C = 0$$

For the first part use F(1) = 0. One is the lower value of the given range for the first part.

$$C = -\frac{1}{5}$$

If $2 \leqslant x \leqslant 4$

$$F(x) = \int \frac{1}{5}(x-1)dx$$
$$= \frac{x^2}{10} - \frac{x}{5} + D$$
$$1 = \frac{4^2}{10} - \frac{4}{5} + D$$
$$D = \frac{1}{5}$$

For the second part use F(4) = 1. Four is the upper value of the range given for the last part.

$$F(x) = \begin{cases} 0, & x \leqslant 1, \\ \frac{1}{5}x - \frac{1}{5}, & 1 < x < 2, \\ \frac{x^2}{10} - \frac{x}{5} + \frac{1}{5} & 2 \leqslant x \leqslant 4, \\ 1. & x > 4. \end{cases}$$

Remember to write the cumulative distribution in full.

Example 6

The random variable X has cumulative distribution function

$$F(x) = \begin{cases} 0, & x < 0, \\ \frac{1}{5}x + \frac{3}{20}x^2, & 0 \leqslant x \leqslant 2, \\ 1, & x > 2 \end{cases}$$

a Find $P(X \leqslant 1.5)$.

b Find $P(0.5 \leqslant X \leqslant 1.5)$.

c Find $P(X = 1)$.

d Find the probability density function, $f(x)$.

a $P(X \leqslant 1.5) = F(1.5)$

$$= \frac{1}{5} \times 1.5 + \frac{3}{20} \times 1.5^2$$
$$= 0.6375$$

Using $F(x) = P(X \leqslant x)$.

b $P(0.5 \leqslant X \leqslant 1.5) = F(1.5) - F(0.5)$

$$= 0.6375 - 0.1375$$
$$= 0.5$$

$P(0.5 \leqslant X \leqslant 1.5) = P(X \leqslant 1.5) - P(X \leqslant 0.5)$.

c $P(X = 1) = 0$

The probability of a single value happening in a continuous distribution is always 0.

d Differentiating $F(x)$ gives $\frac{1}{5} + \frac{3}{10}x$

$$f(x) = \begin{cases} \frac{1}{5} + \frac{3}{10}x, & 0 \leqslant x \leqslant 2, \\ 0. & \text{otherwise.} \end{cases}$$

Exercise 3B

1 The continuous random variable X has probability density function given by:

$$f(x) = \begin{cases} \frac{3x^2}{8}, & 0 \leqslant x \leqslant 2, \\ 0, & \text{otherwise.} \end{cases}$$

Find $F(x)$.

2 The continuous random variable X has probability density function given by:

$$f(x) = \begin{cases} \frac{1}{4}(4 - x), & 1 \leqslant x \leqslant 3, \\ 0, & \text{otherwise.} \end{cases}$$

Find $F(x)$.

3 The continuous random variable X has probability density function given by:

$$f(x) = \begin{cases} \frac{x}{9}, & 0 < x < 3, \\ \frac{1}{9}(6 - x) & 3 \leqslant x \leqslant 6, \\ 0, & \text{otherwise.} \end{cases}$$

Find $F(x)$.

4 The continuous random variable X has probability density function given by:

$$f(x) = \begin{cases} k, & 0 \leqslant x < 3, \\ k(2x - 5), & 3 \leqslant x \leqslant 5 \\ 0, & \text{otherwise.} \end{cases}$$

a Sketch $f(x)$.

b Find the value of k.

c Find $F(x)$.

5 The continuous random variable X has cumulative distribution function given by:

$$F(x) = \begin{cases} 0, & x < 2, \\ \frac{1}{5}(x^2 - 4), & 2 \leqslant x \leqslant 3, \\ 1, & x > 3. \end{cases}$$

Find the probability density function, $f(x)$.

6 The continuous random variable X has cumulative distribution function given by:

$$F(x) = \begin{cases} 0, & x < 1, \\ \frac{1}{2}(x - 1), & 1 \leqslant x \leqslant 3, \\ 1, & x > 3. \end{cases}$$

a Find $P(X \leqslant 2.5)$.

b Find $P(X > 1.5)$.

c Find $P(1.5 \leqslant X \leqslant 2.5)$.

7 The continuous random variable X has probability density function given by:

$$f(x) = \begin{cases} \frac{3x^2}{8} & 0 \leqslant x < 2, \\ 0, & \text{otherwise.} \end{cases}$$

a Find the cumulative distribution function of X.

b Find $P(X \leqslant 1)$.

8 The continuous random variable X has cumulative distribution function given by:

$$F(x) = \begin{cases} 0, & x < 1, \\ \frac{1}{2}(x^3 - 2x^2 + x), & 1 \leqslant x \leqslant 2, \\ 1, & x > 2. \end{cases}$$

a Find the probability density function $f(x)$.

b Sketch the probability density function.

c Find $P(X < 1.5)$.

9 The continuous random variable X has probability density function given by:

$$f(x) = \begin{cases} k(4 - x^2), & 0 \leqslant x \leqslant 2, \\ 0, & \text{otherwise.} \end{cases}$$

a Show that $k = \frac{3}{16}$.

b Find the cumulative distribution function of X.

c Find $P(0.69 < X < 0.70)$. Give your answer correct to one significant figure.

3.3 The mean and variance of a probability density function.

If X is a continuous random variable with p.d.f. $f(x)$:

- **Mean = $\mu = \text{E}(X) = \int_{-\infty}^{\infty} x\text{f}(x)\text{d}x$.**

In book S1, Chapter 8, the mean for a discrete distribution is $\sum x\text{p}(x)$. For a continuous distribution you replace the $\sum$ by $\int_{-\infty}^{\infty}$ and $\text{p}(x)$ by $\text{f}(x)\text{d}x$.

- **Variance = $\text{E}(X^2) - [\text{E}(X)]^2$**

$$= \int_{-\infty}^{\infty} x^2\text{f}(x)\text{d}x - \mu^2$$

The variance for a discrete distribution is $\sum x^2\text{p}(x) - \mu^2$. For a continuous distribution you replace the $\sum$ by $\int_{-\infty}^{\infty}$ and $\text{p}(x)$ by $\text{f}(x)\text{d}x$.
Note $\text{E}(X^2) = \int_{-\infty}^{\infty} x^2\text{f}(x)\text{d}x$.

Example 7

A random variable Y has probability density function

$$\text{f}(y) = \begin{cases} \frac{y}{4}, & 1 \leqslant y \leqslant 3, \\ 0, & \text{otherwise.} \end{cases}$$

Find

a $\text{E}(Y)$, **b** $\text{Var}(Y)$, **c** $\text{E}(2Y - 3)$, **d** $\text{Var}(2Y - 3)$.

a $\text{E}(Y) = \int_1^3 \frac{1}{4}y^2\text{d}y$

$yf(y) = y \times \frac{1}{4}y$.

$$= \left[\frac{1}{12}y^3\right]_1^3$$

$$= \frac{27}{12} - \frac{1}{12}$$

$$= \frac{26}{12}$$

$$= \frac{13}{6}$$

If an exact answer is required you must leave your answer as a fraction. Otherwise you may write the answer as a fraction or as a decimal to 3 significant figures.

b $\text{Var}(Y) = \int_1^3 \frac{1}{4}y^3\text{d}y - \left(\frac{13}{6}\right)^2$

$y^2f(y) = y^2 \times \frac{1}{4}y$.

$$= \left[\frac{1}{16}y^4\right]_1^3 - \left(\frac{13}{6}\right)^2$$

$$= \frac{81}{16} - \frac{1}{16} - \frac{169}{36}$$

$$= \frac{11}{36}$$

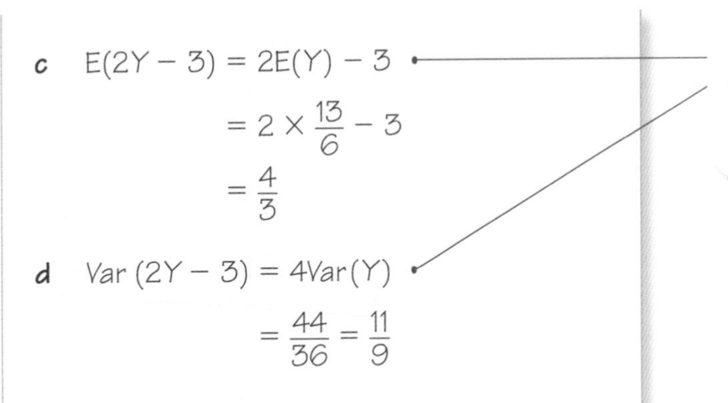

c $E(2Y - 3) = 2E(Y) - 3$

$= 2 \times \frac{13}{6} - 3$

$= \frac{4}{3}$

d $\text{Var}(2Y - 3) = 4\text{Var}(Y)$

$= \frac{44}{36} = \frac{11}{9}$

In book S1 you learnt that
$E(aX + b) = aE(X) + b$,
$\text{Var}(aX + b) = a^2\text{Var}(X)$.
These formulae also apply to continuous variables.

Example 8

A random variable X has probability density function

$$f(x) = \begin{cases} \frac{3}{32}[3 + 2x - x^2], & -1 \leq x \leq 3, \\ 0, & \text{otherwise.} \end{cases}$$

a Sketch the probability density function.

Find

b $E(X)$,

c $P(X > \mu)$.

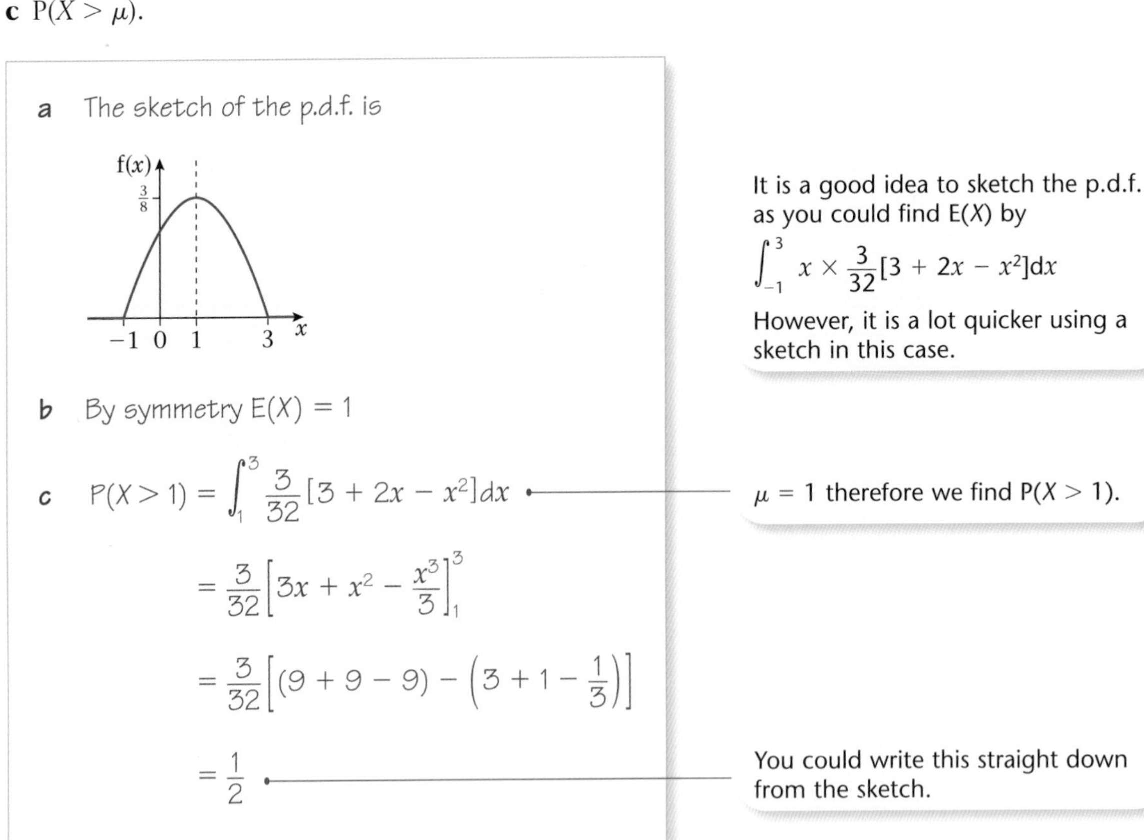

a The sketch of the p.d.f. is

It is a good idea to sketch the p.d.f. as you could find E(X) by

$$\int_{-1}^{3} x \times \frac{3}{32}[3 + 2x - x^2]dx$$

However, it is a lot quicker using a sketch in this case.

b By symmetry $E(X) = 1$

c $P(X > 1) = \int_1^3 \frac{3}{32}[3 + 2x - x^2]dx$

$\mu = 1$ therefore we find $P(X > 1)$.

$= \frac{3}{32}\left[3x + x^2 - \frac{x^3}{3}\right]_1^3$

$= \frac{3}{32}\left[(9 + 9 - 9) - \left(3 + 1 - \frac{1}{3}\right)\right]$

$= \frac{1}{2}$

You could write this straight down from the sketch.

Example 9

The random variable X has probability density function

$$f(x) = \begin{cases} \dfrac{2x}{15}, & 0 \leqslant x < 3, \\ \dfrac{1}{5}(5 - x), & 3 \leqslant x \leqslant 5, \\ 0, & \text{otherwise.} \end{cases}$$

Find

a E(X),

b Var (X).

a $$\mathrm{E}(X) = \int_0^3 \frac{2}{15}x^2 dx + \int_3^5 \frac{1}{5}(5x - x^2)dx$$

$$= \left[\frac{2}{45}x^3\right]_0^3 + \left[\frac{1}{5}\left(\frac{5}{2}x^2 - \frac{x^3}{3}\right)\right]_3^5$$

$$= \left(\frac{6}{5} - 0\right) + \left[\left(\frac{25}{2} - \frac{25}{3}\right) - \left(\frac{9}{2} - \frac{9}{5}\right)\right]$$

$$= 2\frac{2}{3}$$

b $$\mathrm{Var}(X) = \int_0^3 \frac{2}{15}x^3 dx + \int_3^5 \frac{1}{5}(5x^2 - x^3)dx - \left(2\frac{2}{3}\right)^2$$

$$= \left[\frac{2}{60}x^4\right]_0^3 + \left[\frac{1}{5}\left(\frac{5}{3}x^3 - \frac{x^4}{4}\right)\right]_3^5 - \frac{64}{9}$$

$$= \left(\frac{27}{10} - 0\right) + \left[\left(\frac{125}{3} - \frac{125}{4}\right) - \left(9 - \frac{81}{20}\right)\right] - \frac{64}{9}$$

$$= 1\frac{1}{18}$$

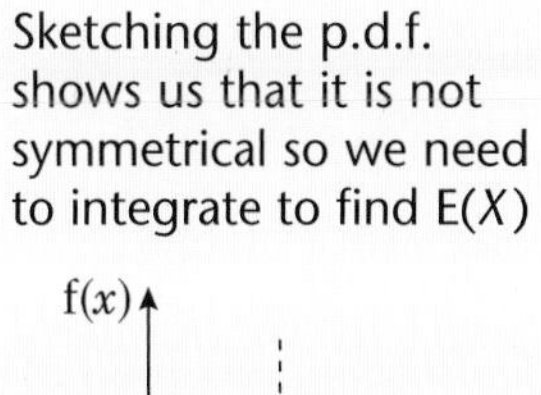

Sketching the p.d.f. shows us that it is not symmetrical so we need to integrate to find E(X)

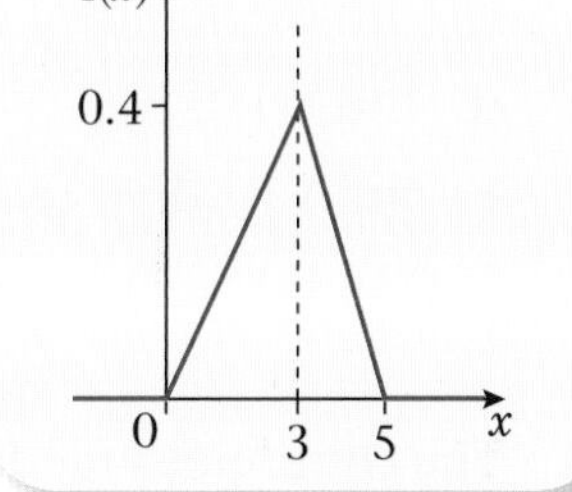

Exercise 3C

1 The continuous random variable X has a probability density function given by

$$f(x) = \begin{cases} kx^2, & 0 \leqslant x \leqslant 2, \\ 0, & \text{otherwise.} \end{cases}$$

Find

a k,

b E(X),

c Var (X).

2 The continuous random variable Y has a probability density function given by

$$f(y) = \begin{cases} \dfrac{y^2}{9}, & 0 \leqslant y \leqslant 3, \\ 0, & \text{otherwise.} \end{cases}$$

a Find $E(Y)$.

b Find $\text{Var}(Y)$.

c Find the standard deviation of Y.

3 The continuous random variable Y has a probability density function given by

$$f(y) = \begin{cases} \dfrac{y}{8}, & 0 \leqslant y \leqslant 4, \\ 0, & \text{otherwise.} \end{cases}$$

a Find $E(Y)$.

b Find $\text{Var}(Y)$.

c Find the standard deviation of Y.

d Find $P(Y > \mu)$.

e Find $\text{Var}(3Y + 2)$.

f Find $E(Y + 2)$.

4 The continuous random variable X has a probability density function given by

$$f(x) = \begin{cases} k(1 - x), & 0 \leqslant x \leqslant 1, \\ 0, & \text{otherwise.} \end{cases}$$

a Find k.

b Find $E(X)$.

c Show that $\text{Var}(X) = \dfrac{1}{18}$.

d Find $P(X > \mu)$.

5 The continuous random variable X has a probability density function given by

$$f(x) = \begin{cases} 12x^2(1 - x), & 0 \leqslant x \leqslant 1, \\ 0, & \text{otherwise.} \end{cases}$$

a Find $P(X < 0.5)$.

b Find $E(X)$.

6 The continuous random variable X has a probability density function given by

$$f(x) = \begin{cases} \dfrac{3}{8}(1 + x^2), & -1 \leqslant x \leqslant 1, \\ 0, & \text{otherwise.} \end{cases}$$

a Sketch the p.d.f. of X.

b Write down $E(X)$.

c Show that $\sigma^2 = 0.4$.

d Find $P(-\sigma < X < \sigma)$.

7 The continuous random variable T has p.d.f. given by

$$f(t) = \begin{cases} kt^3, & 0 \leqslant t \leqslant 2, \\ 0, & \text{otherwise,} \end{cases}$$

where k is a positive constant.

a Find k.

b Show that $E(T)$ is 1.6.

c Find $E(2T + 3)$.

d Find $\text{Var}(T)$.

e Find $\text{Var}(2T + 3)$.

f Find $P(T < 1)$.

8 The continuous random variable X has a probability density function given by

$$f(x) = \begin{cases} \frac{x^2}{27}, & 0 \leqslant x < 3, \\ \frac{1}{3}, & 3 \leqslant x \leqslant 5, \\ 0, & \text{otherwise.} \end{cases}$$

a Draw a rough sketch of $f(x)$.

b Find $E(X)$.

c Find $\text{Var}(X)$

d Find the standard deviation, σ, of X.

9 The continuous random variable X has a probability density function given by

$$f(x) = \begin{cases} \frac{1}{2}(x - 1), & 1 \leqslant x < 2, \\ \frac{1}{6}(5 - x), & 2 \leqslant x \leqslant 5, \\ 0, & \text{otherwise.} \end{cases}$$

a Sketch $f(x)$.

b Find $E(X)$.

c Find $\text{Var}(X)$.

10 Telephone calls arriving at a company are referred immediately by the telephonist to other people working in the company. The time a call takes, in minutes, is modelled by a continuous random variable T, having a p.d.f. given by

$$f(t) = \begin{cases} kt^2, & 0 \leqslant t \leqslant 10, \\ 0, & \text{otherwise.} \end{cases}$$

a Show that $k = 0.003$.

b Find $E(T)$.

c Find $\text{Var}(T)$.

d Find the probability of a call lasting between 7 and 9 minutes.

e Sketch the p.d.f.

3.4 Finding the mode, median and quartiles of a continuous random variable.

■ The **mode** is the value of the random variable where it is most dense i.e. where the p.d.f. reaches its highest point

If X is a continuous random variable with c.d.f. $F(x)$ then the

■ **median, m,** of X is given by $\mathbf{F(m) = 0.5}$,

■ **lower quartile, Q_1,** is given by $\mathbf{F(Q_1) = 0.25}$,

■ **upper quartile, Q_3,** is given by $\mathbf{F(Q_3) = 0.75}$.

Example 10

The random variables X and Y have probability density functions $f(x)$ and $g(y)$ respectively.

$$f(x) = \begin{cases} 12x^2(1 - x), & 0 \leqslant x \leqslant 1, \\ 0, & \text{otherwise.} \end{cases}$$

$$g(y) = \begin{cases} 2y, & 0 \leqslant y \leqslant 1, \\ 0, & \text{otherwise.} \end{cases}$$

Find the mode of

a $f(x)$,

b $g(y)$.

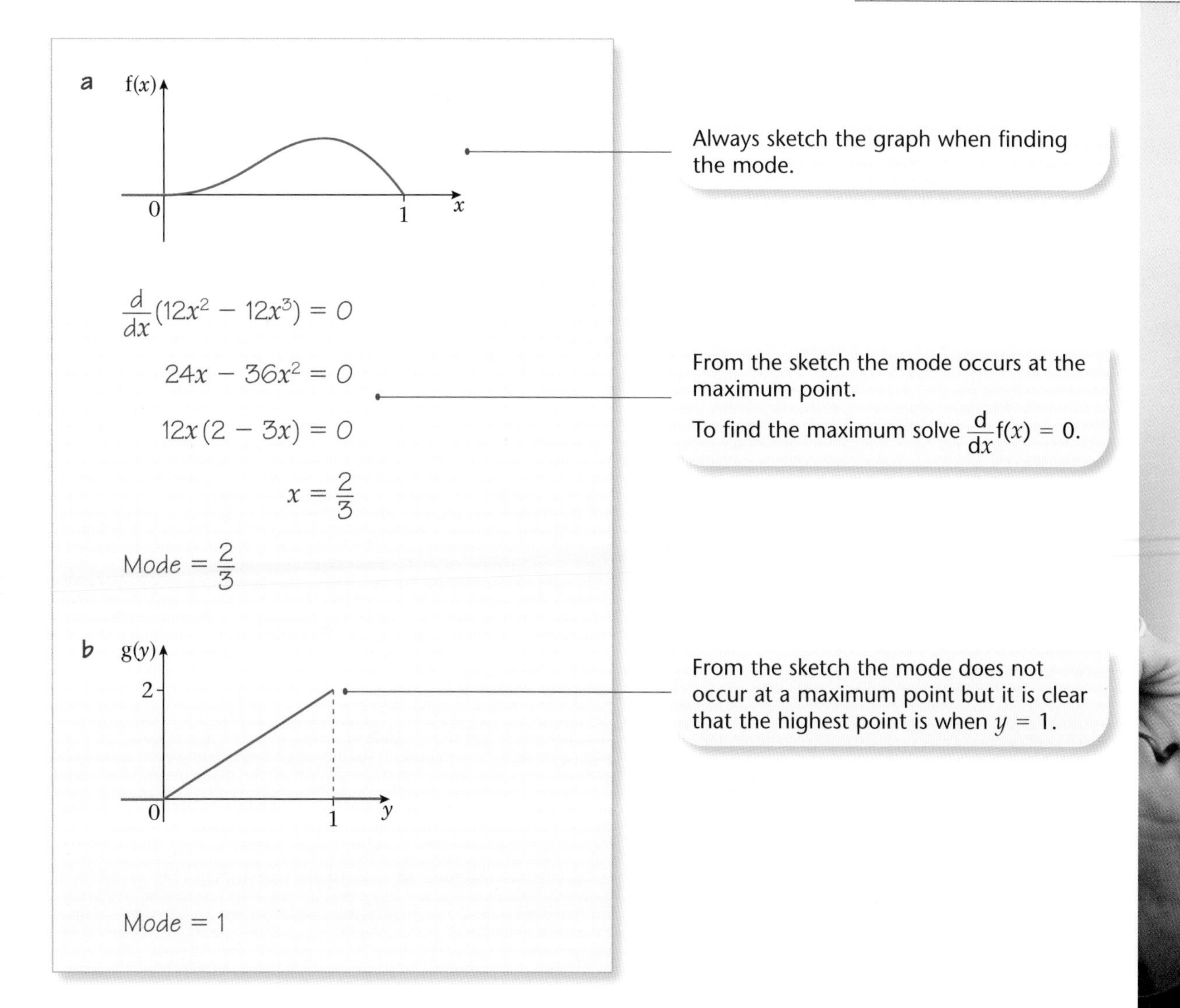

Example 11

A continuous random variable X has probability density function

$$f(x) = \begin{cases} 4x - 4x^3, & 0 \leqslant x \leqslant 1, \\ 0, & \text{otherwise.} \end{cases}$$

Find

a the c.d.f. of X,

b the median value of X.

a Method 1

$$F(x) = \int_0^x 4t - 4t^3 dt$$

$F(x) = \int_0^x f(t)dt$.

$$= [2t^2 - t^4]_0^x$$

$$= 2x^2 - x^4$$

Method 2

$$F(x) = \int 4x - 4x^3 dx$$

$$= 2x^2 - x^4 + C$$

$$F(0) = 0$$

$$C = 0$$

$$F(x) = \begin{cases} 0, & x < 0, \\ 2x^2 - x^4, & 0 \leqslant x \leqslant 1, \\ 1, & x > 1. \end{cases}$$

b $2m^2 - m^4 = 0.5$

$F(x) = 0.5$.

$2m^4 - 4m^2 + 1 = 0$

This is a quadratic equation in m^2.

$$m^2 = \frac{4 \pm \sqrt{16 - 8}}{4}$$

Use the quadratic formula $\frac{-b \pm \sqrt{b^2 - 4ac}}{2a}$.

$$m^2 = 1 \pm \frac{\sqrt{2}}{2}$$

$$m = \sqrt{1 \pm \frac{\sqrt{2}}{2}}$$

$$= 1.31 \text{ or } 0.541$$

Select the value that is in the range of F(x) i.e. $0 \leqslant x \leqslant 1$.

Example 12

A continuous random variable X has the cumulative distribution function

$$F(x) = \begin{cases} 0, & x < 1, \\ \frac{1}{5}x - \frac{1}{5}, & 1 \leqslant x \leqslant 2, \\ \frac{x^2}{10} - \frac{x}{5} + \frac{1}{5}, & 2 < x < 4, \\ 1, & x \geqslant 4. \end{cases}$$

Find the inter-quartile range.

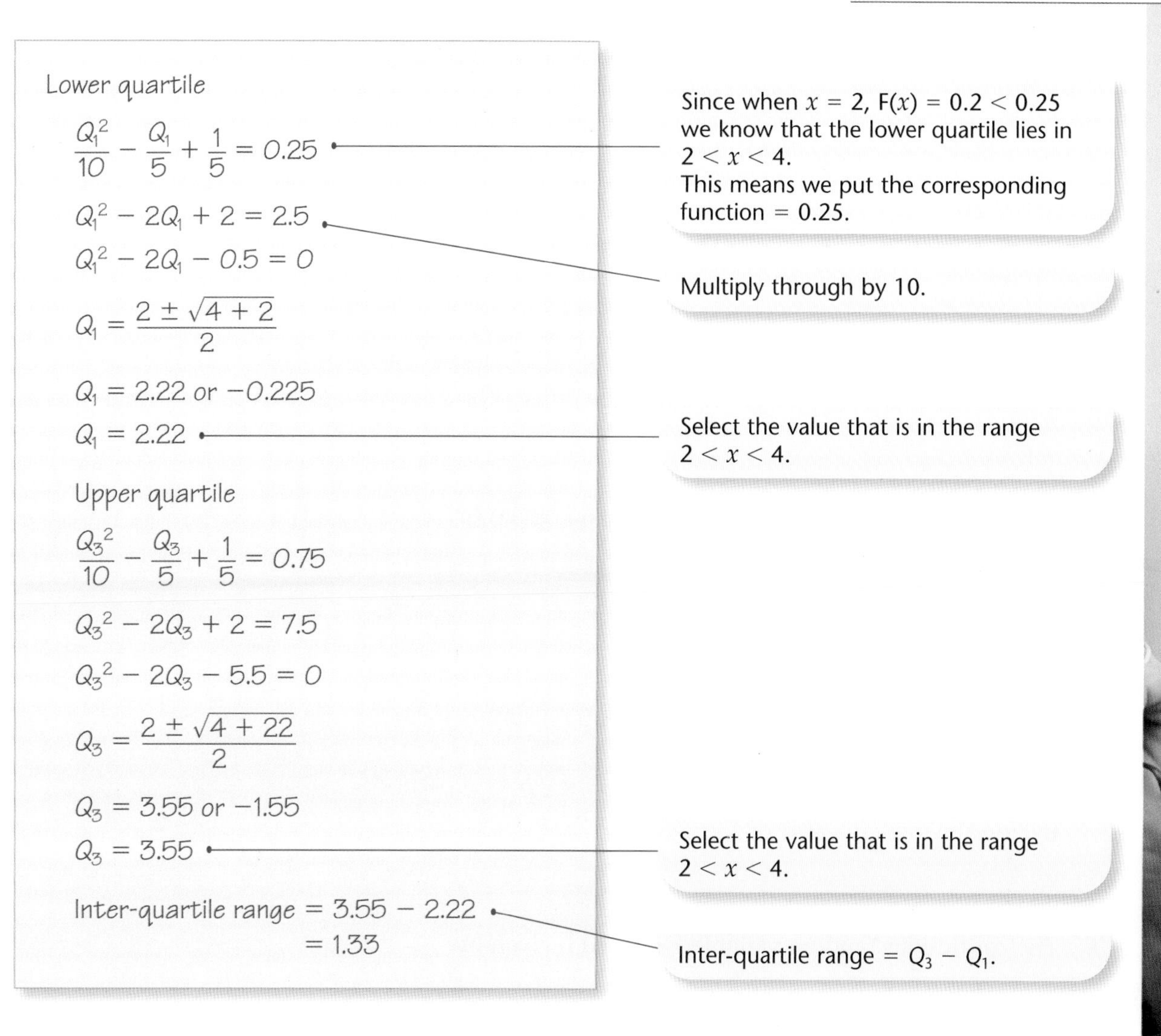

Lower quartile

$$\frac{Q_1^2}{10} - \frac{Q_1}{5} + \frac{1}{5} = 0.25$$

Since when $x = 2$, $F(x) = 0.2 < 0.25$ we know that the lower quartile lies in $2 < x < 4$.
This means we put the corresponding function $= 0.25$.

$$Q_1^2 - 2Q_1 + 2 = 2.5$$

Multiply through by 10.

$$Q_1^2 - 2Q_1 - 0.5 = 0$$

$$Q_1 = \frac{2 \pm \sqrt{4 + 2}}{2}$$

$$Q_1 = 2.22 \text{ or } -0.225$$

$$Q_1 = 2.22$$

Select the value that is in the range $2 < x < 4$.

Upper quartile

$$\frac{Q_3^2}{10} - \frac{Q_3}{5} + \frac{1}{5} = 0.75$$

$$Q_3^2 - 2Q_3 + 2 = 7.5$$

$$Q_3^2 - 2Q_3 - 5.5 = 0$$

$$Q_3 = \frac{2 \pm \sqrt{4 + 22}}{2}$$

$$Q_3 = 3.55 \text{ or } -1.55$$

$$Q_3 = 3.55$$

Select the value that is in the range $2 < x < 4$.

Inter-quartile range $= 3.55 - 2.22$
$= 1.33$

Inter-quartile range $= Q_3 - Q_1$.

Exercise 3D

1 The continuous random variable X has p.d.f. given by

$$f(x) = \begin{cases} \frac{3}{80}(8 + 2x - x^2), & 0 \leqslant x \leqslant 4, \\ 0, & \text{otherwise.} \end{cases}$$

a Sketch the p.d.f. of X.

b Find the mode of X.

2 The continuous random variable X has p.d.f. given by

$$f(x) = \begin{cases} \frac{1}{8}x, & 0 \leqslant x \leqslant 4, \\ 0, & \text{otherwise.} \end{cases}$$

a Sketch the c.d.f. of X.

b Find the median of X.

3 The continuous random variable X has c.d.f. given by

$$F(x) = \begin{cases} 0, & x < 0 \\ \dfrac{x^2}{6}, & 0 \leqslant x < 2, \\ -\dfrac{x^2}{3} + 2x - 2, & 2 \leqslant x \leqslant 3, \\ 1, & x > 3. \end{cases}$$

a Find the median value of X. Give your answer to 3 decimal places.

b Find the quartiles and the inter-quartile range of X. Give your answer to 3 decimal places.

4 The continuous random variable X has p.d.f. given by

$$f(x) = \begin{cases} 1 - \dfrac{1}{2}x, & 0 \leqslant x \leqslant 2, \\ 0, & \text{otherwise.} \end{cases}$$

a Sketch the p.d.f. of X.

b Write down the mode of X.

c Find the c.d.f. of X.

d Find the median value of X.

5 The continuous random variable Y has p.d.f. given by

$$f(y) = \begin{cases} \dfrac{1}{2} - \dfrac{1}{9}y, & 0 \leqslant y \leqslant 3, \\ 0, & \text{otherwise.} \end{cases}$$

a Sketch the p.d.f. of Y.

b Write down the mode of Y.

c Find the c.d.f. of Y.

d Find the median value of Y.

6 The continuous random variable X has p.d.f. given by

$$f(x) = \begin{cases} \dfrac{1}{4}x^3, & 0 \leqslant x \leqslant 2, \\ 0, & \text{otherwise.} \end{cases}$$

a Sketch the p.d.f. of X.

b Write down the mode of X.

c Find the c.d.f. of X.

d Find the median value of X.

7 The continuous random variable X has p.d.f. given by

$$f(x) = \begin{cases} \frac{3}{8}(x^2 + 1), & -1 \leqslant x \leqslant 1, \\ 0, & \text{otherwise.} \end{cases}$$

a Sketch the p.d.f. of X.

b What can you say about the mode of X?

c Write down the median value of X.

d Find the c.d.f. of X.

8 The continuous random variable X has p.d.f. given by

$$f(x) = \begin{cases} \frac{3}{10}(3x - x^2), & 0 \leqslant x \leqslant 2, \\ 0, & \text{otherwise.} \end{cases}$$

a Sketch the p.d.f. of X.

b Find the mode of X.

c Find the c.d.f. of X.

d Show that the median value of X lies between 1.23 and 1.24.

9 The continuous random variable X has c.d.f. given by

$$F(x) = \begin{cases} 0, & x < 1, \\ \frac{1}{8}(x^2 - 1), & 1 \leqslant x \leqslant 3, \\ 1, & x > 3. \end{cases}$$

a Find the p.d.f. of the random variable X.

b Find the mode of X.

c Find the median of X.

d Find the quartiles of X.

10 The continuous random variable X has c.d.f. given by

$$F(x) = \begin{cases} 0, & x < 0, \\ 4x^3 - 3x^4, & 0 \leqslant x \leqslant 1, \\ 1, & x > 1. \end{cases}$$

a Find the p.d.f. of the random variable X.

b Find the mode of X.

c Find $P(0.2 < X < 0.5)$.

11 The amount of vegetables eaten by a family in a week is a continuous random variable W kg. The continuous random variable W has p.d.f. given by

$$f(w) = \begin{cases} \frac{20}{5^5} w^3(5 - w), & 0 \leqslant w \leqslant 5, \\ 0, & \text{otherwise.} \end{cases}$$

a Find the c.d.f. of the random variable W.

b Find, to 3 decimal places, the probability that the family eat between 2 kg and 4 kg of vegetables in one week. **E**

12 The continuous random variable X has a probability density function given by

$$f(x) = \begin{cases} \frac{1}{4}, & 0 \leqslant x < 1, \\ \frac{x^3}{5}, & 1 \leqslant x \leqslant 2, \\ 0, & \text{otherwise.} \end{cases}$$

a Find the cumulative distribution function.

b Find, to 3 decimal places, the median and the inter-quartile range of the distribution. **E**

Mixed exercise 3E

1 The random variable X has probability density function f(x) given by

$$f(x) = \begin{cases} \frac{1}{3}\left(1 + \frac{x}{2}\right), & 0 \leqslant x \leqslant 2, \\ 0, & \text{otherwise.} \end{cases}$$

Find

a E(X) and E($3X + 2$),

b Var (X) and Var ($3X + 2$),

c $P(X < 1)$,

d $P(X > \mu)$,

e $P(0.5 < X < 1.5)$.

2 The random variable X has probability density function f(x) given by

$$f(x) = \begin{cases} 2 - 2x, & 0 \leqslant x \leqslant 1, \\ 0, & \text{otherwise.} \end{cases}$$

a Evaluate E(X).

b Evaluate Var (X).

c Write down the values of E($2X + 1$) and Var ($2X + 1$).

d Specify fully the cumulative distribution function of X.

e Work out the median value of X.

3 The continuous random variable Y has cumulative distribution function given by

$$\mathrm{F}(y) = \begin{cases} 0, & y < 1, \\ k(y^2 - y), & 1 \leqslant y \leqslant 2, \\ 1, & y > 2. \end{cases}$$

where k is a positive constant.

a Show that $k = \frac{1}{2}$.

b Find $\mathrm{P}(Y < 1.5)$.

c Find the value of the median.

d Specify fully the probability density function $\mathrm{f}(y)$.

4 The continuous random variable X has cumulative distribution function

$$\mathrm{F}(x) = \begin{cases} 0, & x < 2, \\ \frac{1}{5}(x^2 - 4), & 2 \leqslant x \leqslant 3, \\ 1, & x > 3. \end{cases}$$

a Find $\mathrm{P}(X > 2.4)$.

b Find the median.

c Find the probability density function, $\mathrm{f}(x)$.

d Evaluate $\mathrm{E}(X)$.

e Find the mode of X.

5 The random variable X has probability density function $\mathrm{f}(x)$ given by

$$\mathrm{f}(x) = \begin{cases} kx^2, & 0 \leqslant x \leqslant 2, \\ 0, & \text{otherwise.} \end{cases}$$

where k is a positive constant.

a Show that $k = \frac{3}{8}$.

b Calculate $\mathrm{E}(X)$.

c Specify fully the cumulative distribution function of X.

d Find the value of the median.

e Find the value of the mode.

6 The random variable Y has probability density function $\mathrm{f}(y)$ given by

$$\mathrm{f}(y) = \begin{cases} k(y^2 + 2y + 2), & 1 \leqslant y \leqslant 3, \\ 0, & \text{otherwise.} \end{cases}$$

where k is a positive constant.

a Show that $k = \frac{3}{62}$.

b Specify fully the cumulative distribution function of Y.

c Evaluate $\mathrm{P}(Y \leqslant 2)$.

7 A random variable X has probability density function f(x) given by

$$f(x) = \begin{cases} \frac{3}{32}(4 - x^2), & -2 \leqslant x \leqslant 2, \\ 0, & \text{otherwise.} \end{cases}$$

a Sketch the probability density function of X.

b Write down the mode of X.

c Specify fully the cumulative distribution function of X.

d Find $P(0.5 < X < 1.5)$.

8 A random variable X has probability density function f(x) given by

$$f(x) = \begin{cases} \frac{1}{3}, & 0 \leqslant x < 1, \\ \frac{2}{7}x^2, & 1 \leqslant x \leqslant 2, \\ 0, & \text{otherwise.} \end{cases}$$

a Find E(X).

b Specify fully the cumulative distribution function of X.

c Find the median of X.

9 A continuous random variable X has probability density function f(x) given by

$$f(x) = \begin{cases} kx - k, & 1 \leqslant x \leqslant 3, \\ 0, & \text{otherwise.} \end{cases}$$

where k is a positive constant.

a Show that $k = \frac{1}{2}$.

b Find E(X).

c Work out the cumulative distribution function, F(x).

d Show that the median value lies between 2.4 and 2.5.

10 The continuous random variable X has probability density function given by

$$f(x) = \begin{cases} x, & 0 \leqslant x < 1, \\ \frac{3x^2}{14}, & 1 \leqslant x \leqslant 2, \\ 0, & \text{otherwise.} \end{cases}$$

a Sketch the probability density function of X.

b Find the mode of X.

c Find E($2X$).

d Find Var($2X + 1$)

e Specify fully the cumulative distribution function of X.

f Using your answer to part **e** find the median of X.

Summary of key points

1 For a continuous random variable, X

$$\int_{-\infty}^{\infty} \mathrm{f}(x)\,\mathrm{d}x = 1$$

$$\mu = \mathrm{E}(X) = \int_{-\infty}^{\infty} x\mathrm{f}(x)\,\mathrm{d}x$$

$$\sigma^2 = \mathrm{E}(X^2) - \mu^2 = \int_{-\infty}^{\infty} x^2\mathrm{f}(x)\,\mathrm{d}x - \mu^2$$

2 Cumulative distribution function, $\mathrm{F}(x)$

$$0 \leqslant \mathrm{F}(x) \leqslant 1$$

$$\mathrm{F}(x) = \mathrm{P}(X < x) = \int_{-\infty}^{x} \mathrm{f}(t)\,\mathrm{d}t$$

3 The median m satisfies $\mathrm{F}(m) = 0.5$.
The Lower Quartile Q_1 satisfies $\mathrm{F}(Q_1) = 0.25$.
The Upper Quartile Q_3 satisfies $\mathrm{F}(Q_3) = 0.75$.

4 The mode is the x value at the highest point of the function.

1 Review Exercise

1 It is estimated that 4% of people have green eyes. In a random sample of size n, the expected number of people with green eyes is 5.

a Calculate the value of n.

The expected number of people with green eyes in a second random sample is 3.

b Find the standard deviation of the number of people with green eyes in this second sample. *E*

2 In a manufacturing process, 2% of the articles produced are defective. A batch of 200 articles is selected.

a Giving a justification for your choice, use a suitable approximation to estimate the probability that there are exactly 5 defective articles.

b Estimate the probability there are less than 5 defective articles. *E*

3 A continuous random variable X has probability density function

$$f(x) = \begin{cases} k(4x - x^3), & 0 \leqslant x \leqslant 2, \\ 0, & \text{otherwise,} \end{cases}$$

where k is a positive constant.

a Show that $k = \frac{1}{4}$.

Find

b $E(X)$,

c the mode of X,

d the median of X.

e Comment on the skewness of the distribution.

f Sketch $f(x)$. *E*

4 A fair coin is tossed 4 times.
Find the probability that

a an equal number of heads and tails occur,

b all the outcomes are the same,

c the first tail occurs on the third throw. *E*

5 Accidents on a particular stretch of motorway occur at an average rate of 1.5 per week.

a Write down a suitable model to represent the number of accidents per week on this stretch of motorway.

Find the probability that

b there will be 2 accidents in the same week,

c there is at least one accident per week for 3 consecutive weeks,

d there are more than 4 accidents in a two-week period. **E**

6 The random variable $X \sim B(150, 0.02)$. Use a suitable approximation to estimate $P(X > 7)$. **E**

7 A continuous random variable X has probability density function $f(x)$ where,

$$f(x) = \begin{cases} kx(x-2), & 2 \leqslant x \leqslant 3, \\ 0, & \text{otherwise,} \end{cases}$$

where k is a positive constant.

a Show that $k = \frac{3}{4}$.

Find

b $E(X)$,

c the cumulative distribution function $F(x)$.

d Show that the median value of X lies between 2.70 and 2.75. **E**

8 The probability of a bolt being faulty is 0.3. Find the probability that in a random sample of 20 bolts there are

a exactly 2 faulty bolts,

b more than 3 faulty bolts.

These bolts are sold in bags of 20. John buys 10 bags.

c Find the probability that exactly 6 of these bags contain more than 3 faulty bolts. **E**

9 **a** State two conditions under which a Poisson distribution is a suitable model to use in statistical work.

The number of cars passing an observation point in a 10-minute interval is modelled by a Poisson distribution with mean 1.

b Find the probability that in a randomly chosen 60-minute period there will be

i exactly 4 cars passing the observation point,

ii at least 5 cars passing the observation point.

The number of other vehicles, (i.e. other than cars), passing the observation point in a 60-minute interval is modelled by a Poisson distribution with mean 12.

c Find the probability that exactly 1 vehicle, of **any type**, passes the observation point in a 10-minute period. **E**

10 The continuous random variable Y has cumulative distribution function $F(y)$ given by

$$F(y) = \begin{cases} 0, & y < 1, \\ k(y^4 + y^2 - 2), & 1 \leqslant y \leqslant 2, \\ 1, & y > 2. \end{cases}$$

a Show that $k = \frac{1}{18}$.

b Find $P(Y > 1.5)$.

c Specify fully the probability density function $f(y)$. **E**

11 The continuous random variable X has probability density function $f(x)$ given by

$$f(x) = \begin{cases} 2(x-2), & 2 \leqslant x \leqslant 3, \\ 0, & \text{otherwise.} \end{cases}$$

a Sketch $f(x)$ for all values of x.

b Write down the mode of X.

Find

c $E(X)$,

d the median of X.

e Comment on the skewness of this distribution. Give a reason for your answer. *E*

12 An engineering company manufactures an electronic component. At the end of the manufacturing process, each component is checked to see if it is faulty.
Faulty components are detected at a rate of 1.5 per hour.

a Suggest a suitable model for the number of faulty components detected per hour.

b Describe, in the context of this question, two assumptions you have made in part **a** for this model to be suitable.

c Find the probability of 2 faulty components being detected in a 1-hour period.

d Find the probability of at least one faulty component being detected in a 3-hour period. *E*

13 **a** Write down the conditions under which the Poisson distribution may be used as an approximation to the binomial distribution.

A call centre routes incoming telephone calls to agents who have specialist knowledge to deal with the call. The probability of the caller being connected to the wrong agent is 0.01.

b Find the probability that 2 consecutive calls will be connected to the wrong agent.

c Find the probability that more than 1 call in 5 consecutive calls are connected to the wrong agent.

The call centre receives 1000 calls each day.

d Find the mean and variance of the number of wrongly connected calls.

e Use a Poisson approximation to find, to 3 decimal places, the probability that more than 6 calls each day are connected to the wrong agent. *E*

14 The continuous random variable X has probability density function given by

$$f(x) = \begin{cases} \frac{1}{6}x, & 0 < x < 3, \\ 2 - \frac{1}{2}x, & 3 \leqslant x < 4, \\ 0, & \text{otherwise.} \end{cases}$$

a Sketch the probability density function of X.

b Find the mode of X.

c Specify fully the cumulative distribution function of X.

d Using your answer to part **c**, find the median of X. *E*

15 The random variable J has a Poisson distribution with mean 4.

a Find $P(J \geqslant 10)$

The random variable K has a binomial distribution with parameters $n = 25$, $p = 0.27$.

b Find $P(K \leqslant 1)$ *E*

16 The continuous random variable X has cumulative distribution function

$$F(x) = \begin{cases} 0, & x < 0, \\ 2x^2 - x^3, & 0 \leqslant x \leqslant 1, \\ 1, & x > 1. \end{cases}$$

a Find $P(X > 0.3)$.

b Verify that the median value of X lies between $x = 0.59$ and $x = 0.60$.

c Find the probability density function $f(x)$.

d Evaluate $E(X)$.

e Find the mode of X.

f Comment on the skewness of X. Justify your answer.

After completing this chapter you should

- be able to decide when to use the continuous uniform distribution
- be able to derive the mean, variance and cumulative distribution function of a continuous uniform distribution
- be able to use the formulae for the mean, variance and cumulative distribution function of a continuous uniform distribution.

Continuous uniform distribution

LEVEL 4 ENERGY TIME: 14:23

In a computer game an alien appears every 2 seconds. The player stops the alien by pressing a key. The object of the game is to stop the alien as soon as it appears. A simple model of the game assumes that the time taken, T s, to stop the alien is a continuous uniform random variable defined over the interval [0, 1].

Find the probability that the alien is stopped within 0.2 seconds of it appearing.

When you have finished this chapter you will be able to answer such questions.

4.1 Continuous uniform/rectangular distribution.

The continuous random variable X with p.d.f.

$$\mathrm{f}(x) = \begin{cases} \dfrac{1}{b-a}, & a \leqslant x \leqslant b, \\ 0, & \text{otherwise.} \end{cases}$$

where a and b are constants is called a **continuous uniform (rectangular) distribution**.

It is denoted by $X \sim \mathbf{U}[\boldsymbol{a}, \boldsymbol{b}]$.

A sketch of the p.d.f. is as follows.

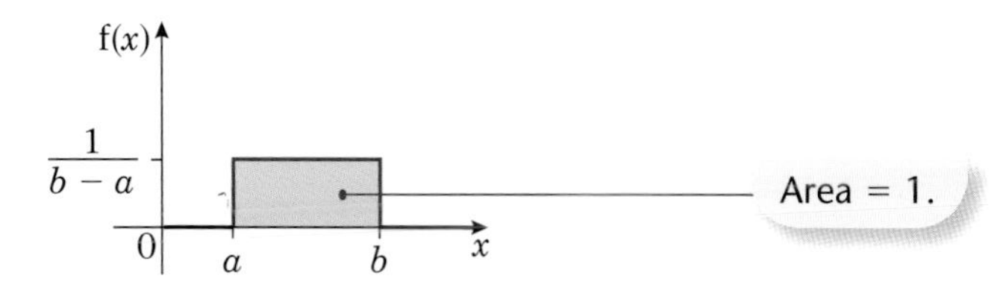

Example 1

The continuous random variable $X \sim \mathrm{U}[3, 5]$

a Write down the distribution of $Y = 5X - 4$.

b Find $\mathrm{P}(3.2 < X < 4.3)$.

a Substitute the lower and upper limits of 3 and 5 into $Y = 5X - 4$ to get

lower limit: $\mathrm{E}(Y) = 5\mathrm{E}(X) - 4$

$= 5 \times 3 - 4$

$= 11$

upper limit: $\mathrm{E}(Y) = 5 \times 5 - 4$

$= 21$

$Y \sim \mathrm{U}[11, 21]$

Remember: $\mathrm{E}(aX + b) = a\mathrm{E}(X) + b$.

b $\dfrac{1}{b-a} = \dfrac{1}{5-3} = 0.5$

When dealing with a uniform distribution it is easier to sketch the p.d.f. and work out the area of the rectangle

$\mathrm{P}(3.2 < X < 4.3) = (4.3 - 3.2) \times 0.5$

$= 0.55$

$\mathrm{P}(3.2 < x < 4.3)$ is area of the shaded section on the sketch.

Example 2

The continuous random variable X has p.d.f as shown in the diagram.

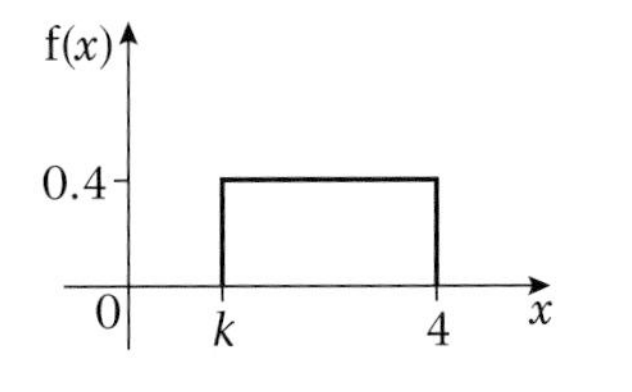

Find

a the value of k,

b $\mathrm{P}(3 < X < 3.5)$.

a	Area = 1
	$0.4 \times (4 - k) = 1$
	$4 - k = 2.5$
	$k = 1.5$
b	$\mathrm{P}(3 < X < 3.5) = 0.4 \times (3.5 - 3) = 0.2$

Exercise 4A

1 The continuous random variable $X \sim \mathrm{U}[2, 7]$.
Find

a $\mathrm{P}(3 < X < 5)$,

b $\mathrm{P}(X > 4)$.

2 The continuous random variable X has p.d.f. as shown in the diagram.

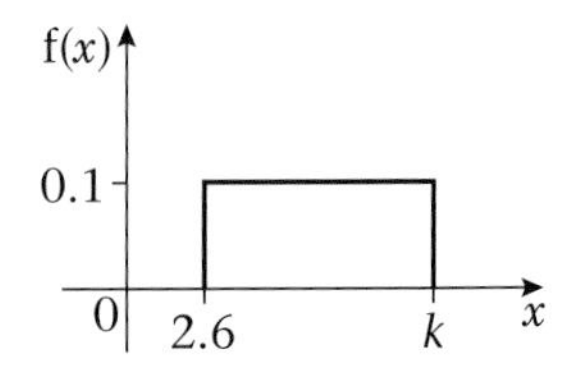

Find

a the value of k,

b $\mathrm{P}(4 < X < 7.9)$.

3 The continuous random variable X has p.d.f.

$$\mathrm{f}(x) = \begin{cases} k, & -2 \leqslant x \leqslant 6, \\ 0, & \text{otherwise.} \end{cases}$$

Find

a the value of k,

b $P(-1.3 < X < 4.2)$.

4 The continuous random variable $Y \sim U[a, b]$. Given that $P(Y < 5) = \frac{1}{4}$ and $P(Y > 7) = \frac{1}{2}$, find the value of a and the value of b.

5 The continuous random variable $X \sim U[2, 8]$.

a Write down the distribution of $Y = 2X + 5$.

b Find $P(12 < Y < 20)$.

4.2 The properties of a continuous uniform distribution.

Example 3

The continuous random variable X has p.d.f.

$$f(x) = \begin{cases} \dfrac{1}{b-a}, & a \leqslant x \leqslant b, \\ 0, & \text{otherwise.} \end{cases}$$

Find

a $E(X)$,

b $Var(X)$,

c $F(x)$.

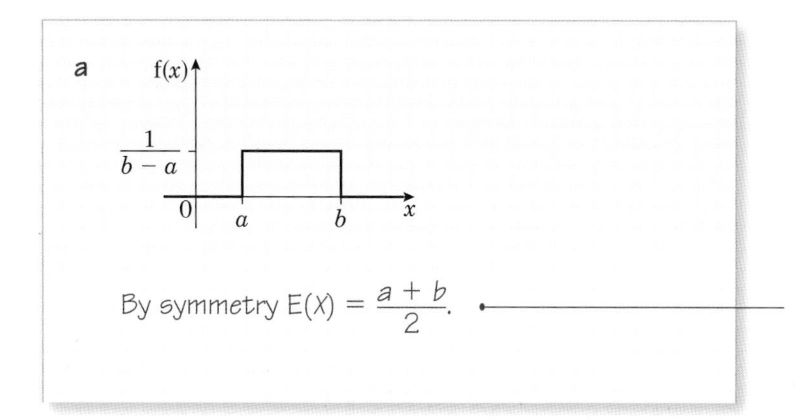

By symmetry $E(X) = \dfrac{a+b}{2}$.

The formulae for E(X) for the continuous uniform distribution is given in the formula book. However, you should be able to derive it from first principles.

b $\text{Var}(X) = \int_a^b \frac{(x-\bar{x})^2}{b-a}\,dx$

This is using the continuous version of Variance $= \frac{\Sigma(x-\bar{x})^2}{n}$.

You could use $E(X^2) - (E(X))^2$ but it is more difficult in this case.

$$= \int_a^b \left(x - \left(\frac{a+b}{2}\right)\right)^2 \times \frac{1}{(b-a)}\,dx$$

$$= \left[\frac{\left(x - \left(\frac{a+b}{2}\right)\right)^3}{3(b-a)}\right]_a^b$$

Using $\int (x+a)^n dx = \frac{(x+a)^{n+1}}{n+1}$

$$= \frac{\left(b - \left(\frac{a+b}{2}\right)\right)^3}{3(b-a)} - \frac{\left(a - \left(\frac{a+b}{2}\right)\right)^3}{3(b-a)}$$

$$= \frac{\left(\frac{b-a}{2}\right)^3 - \left(\frac{a-b}{2}\right)^3}{3(b-a)}$$

$$= \frac{\left(\frac{b-a}{2}\right)^3 + \left(\frac{b-a}{2}\right)^3}{3(b-a)}$$

Using $-(a-b)^3 = (b-a)^3$

$$= \frac{\frac{(b-a)^3}{8} + \frac{(b-a)^3}{8}}{3(b-a)}$$

$$= \frac{\frac{(b-a)^3}{4}}{3(b-a)}$$

$$= \frac{(b-a)^3}{12(b-a)}$$

$$= \frac{(b-a)^2}{12}$$

The formulae for $\text{Var}(X)$ for the continuous uniform distribution is given in the formula book. However, you should be able to derive it from first principles.

c If $a \leqslant x \leqslant b$, $F(x) = \int_a^x \frac{1}{b-a}\,dt$

$$= \left[\frac{t}{b-a}\right]_a^x$$

$$= \frac{x-a}{b-a}$$

$$F(x) = \begin{cases} 0, & x < a, \\ \frac{x-a}{b-a} & a \leqslant x \leqslant b, \\ 1, & x > b. \end{cases}$$

You may learn and then use the formula for $F(x)$ for the uniform continuous distribution. However, you should be able to find $F(x)$ from first principles.

For a continuous uniform distribution $U[a, b]$

- $E(X) = \frac{a+b}{2}$
- $Var(X) = \frac{(b-a)^2}{12}$
- $F(X) = \begin{cases} 0, & x < a \\ \frac{x-a}{b-a}, & a \leqslant x \leqslant b, \\ 1, & x > b. \end{cases}$

Example 4

The continuous random variable Y is uniformly distributed over the interval [4, 7]. Find:

a $E(X)$,

b $Var(X)$,

c the cumulative distribution function of X, for all x.

a $E(X) = \frac{4+7}{2}$

$= 5.5$

b $Var(X) = \frac{(7-4)^2}{12}$

$= \frac{3}{4}$

c $F(x) = \int_4^x \frac{1}{7-4} dx$

$= \left[\frac{x}{3}\right]_4^x$

$= \frac{x-4}{3}$

$F(x) = \begin{cases} 0, & x < 4, \\ \frac{x-4}{3}, & 4 \leqslant x \leqslant 7, \\ 1, & x > 7. \end{cases}$

You can write this down straight away if you learn the formula for $F(X)$ of a continuous uniform distribution.

Example 5

The continuous random variable Y is uniformly distributed over the interval $[a, b]$. Given $\text{E}(Y) = 1$ and $\text{Var}(Y) = \frac{16}{3}$, find the value of a and the value of b.

E(Y)

$$\frac{a+b}{2} = 1$$

$$a + b = 2 \qquad (1)$$

Var(Y)

$$\frac{(b-a)^2}{12} = \frac{16}{3}$$

$$(b-a)^2 = 64 \qquad (2)$$

Solving equations (1) and (2) simultaneously

$$b = 2 - a$$

$$(2 - a - a)^2 = 64$$

$$(2 - 2a) = \pm 8$$

$$2 - 2a = 8 \qquad\qquad 2 - 2a = -8$$

$$a = -3 \qquad\qquad a = 5$$

$$b = 2 - -3 \qquad\qquad b = 2 - 5$$

$$= 5 \qquad\qquad = -3$$

Since $a < b$, $a = -3$ and $b = 5$.

Example 6

The continuous variable X is uniformly distributed over the interval $[-3, 5]$.

a Write down $\text{E}(X)$.

b Use integration to find the Variance of X.

a $\text{E}(X) = 1$

b $\text{Var}(X) = \text{E}(X^2) - [\text{E}(X)]^2$

$$= \int_{-3}^{5} \frac{x^2}{8} dx - 1^2$$

$$= \left[\frac{x^3}{24}\right]_{-3}^{5} - 1$$

$$= \frac{125}{24} + \frac{27}{24} - 1$$

$$= 5\frac{1}{3}$$

You could have done

$$\int_{-3}^{5} \frac{(x-1)^2}{8} = \left[\frac{(x-1)^3}{24}\right]_{-3}^{5}$$

$$= \frac{4^3}{24} - \frac{(-4)^3}{24}$$

$$= 5\frac{1}{3}$$

Example 7

The trunk of a small tree varies in diameter from 10 cm at the bottom to 2 cm at the top. A small child is asked to measure the diameter of the trunk. The random variable R is the radius of the cross section of the tree as measured by the child. $R \sim \text{U}[1, 5]$.

Find the expected value of the area, A, of the cross-section of the tree.

$A = \pi R^2$

$\text{E}(A) = \text{E}(\pi R^2)$

$= \pi\text{E}(R^2)$

$\text{Var}(R) = \text{E}(R^2) - [\text{E}(R)]^2$

Rearranging gives

$\text{E}(R^2) = \text{Var}(R) + [\text{E}(R)]^2$

$\text{Var}(R) = \dfrac{(5-1)^2}{12} = 1\frac{1}{3}$

$\text{E}(R) = \dfrac{5+1}{2} = 3$

$\text{E}(R^2) = 1\frac{1}{3} + 9$

$= 10\frac{1}{3}$

$\text{E}(A) = \dfrac{31\pi}{3}$

To find $\text{E}(R^2)$ you could have used

$\text{E}(X^2) = \int x^2\text{f}(x)\text{d}x$

$\int_1^5 \frac{1}{4}r^2\text{d}r = \left[\frac{1}{12}r^3\right]_1^5$ $\left(\text{since } \frac{1}{b-a} = \frac{1}{5-1} = \frac{1}{4}\right)$

$= \frac{125}{12} - \frac{1}{12}$

$= 10\frac{1}{3}$

Exercise 4B

1 The continuous variable Y is uniformly distributed over the interval $[-3, 5]$. Find:

a $\text{E}(X)$,

b $\text{Var}(X)$,

c $\text{E}(X^2)$,

d the cumulative distribution function of X, for all x.

2 Find $\text{E}(X)$ and $\text{Var}(X)$ for the following probability density functions.

a $\text{f}(x) = \begin{cases} \frac{1}{4}, & 1 \leqslant x \leqslant 5, \\ 0, & \text{otherwise.} \end{cases}$

b $\text{f}(x) = \begin{cases} \frac{1}{8}, & -2 \leqslant x \leqslant 6, \\ 0, & \text{otherwise.} \end{cases}$

3 The continuous random variable X has p.d.f as shown in the diagram.

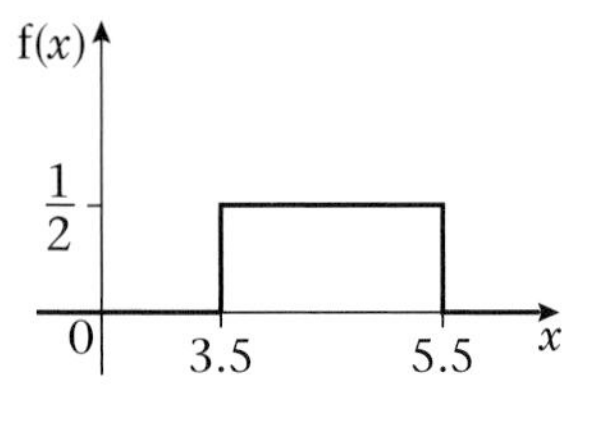

Find:

a $E(X)$,

b $Var(X)$,

c $E(X^2)$,

d the cumulative distribution function of X, for all x.

4 The continuous random variable $Y \sim U[a, b]$. Given $E(Y) = 1$ and $Var(Y) = \frac{4}{3}$, find the value of a and the value of b.

5 The continuous random variable X has probability density function

$$f(x) = \begin{cases} \frac{1}{6}, & -1 \leqslant x \leqslant 5, \\ 0, & \text{otherwise.} \end{cases}$$

Given that $Y = 4X - 6$, find $E(Y)$ and $Var(Y)$.

6 The random variable X is the length of a side of a square. $X \sim U[4.5, 5.5]$.
The random variable Y is the area of the square.
Find $E(Y)$.

7 In a computer game an alien appears every 2 seconds. The player stops the alien by pressing a key. The object of the game is to stop the alien as soon as it appears. Given that the player actually presses the key T s after the alien first appears, a simple model of the game assumes that T is a continuous uniform random variable defined over the interval [0, 1].

a Write down $P(T < 0.2)$.

b Write down $E(T)$.

c Use integration to find $Var(T)$.

4.3 Choosing the right model.

Example 8

The length of a pencil is measured to the nearest cm. Write down the distribution of the rounding errors R.

The error is the difference between the true length and the recorded length.

If a pencil is recorded as 20 cm long then its length is anywhere in the interval

$19.5 \text{ cm} \leqslant \text{length} < 20.5 \text{ cm}$

The error is therefore in the interval

$-0.5 \leqslant \text{error} < 0.5.$

As it is reasonable to assume that the error is equally likely to take any of the values in this range

$R \sim U[-0.5, 0.5]$

The uniform distribution is often used as a model for errors made by rounding up or down when recording measurements.

Example 9

A bus arrives, on time, at a bus stop every 20 minutes. Jimmy arrives at the bus stop at a random time without knowing when the next bus is due. Let X represent the time Jimmy has to wait for a bus to arrive at the bus stop.

a Suggest a suitable model for the distribution of X.

b Using your model calculate the probability that Jimmy will wait more than 6 minutes for a bus to arrive.

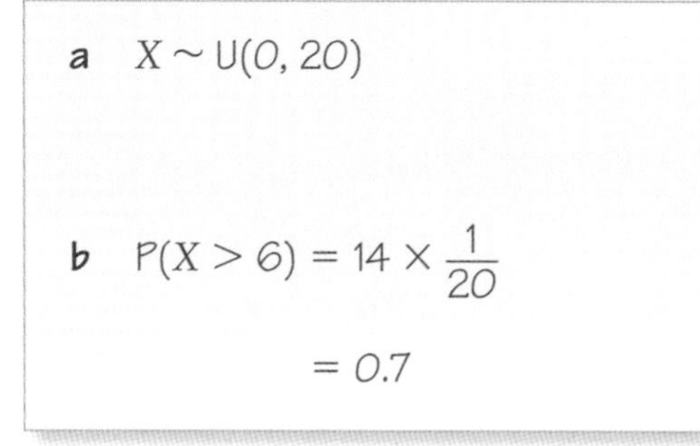

a $X \sim U(0, 20)$

b $P(X > 6) = 14 \times \frac{1}{20}$

$= 0.7$

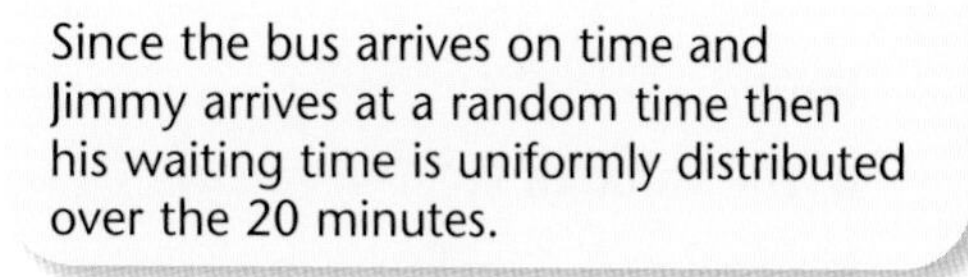

Since the bus arrives on time and Jimmy arrives at a random time then his waiting time is uniformly distributed over the 20 minutes.

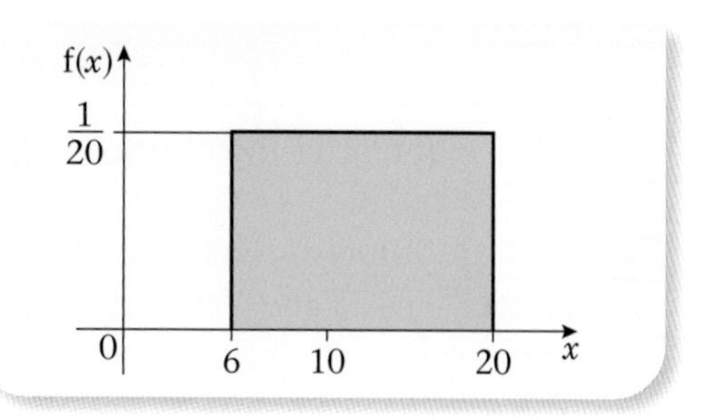

Example 10

Write down the name of the distribution you would recommend as a suitable model for each of the following situations.

a The weights of 200 g tins of tomatoes produced on a production line.

b The difference between the true length and the length of metal rods measured to the nearest centimetre.

a Normal — You expect more tins to be near the 200 g mark.

b Continuous uniform — It is reasonable to assume that the difference is equally likely to take any of the values in the range −0.5 to 0.5.

Mixed exercise 4C

1 The continuous random variable X is uniformly distributed over the interval $[-2, 5]$.

a Sketch the probability density function $f(x)$ of X.

Find

b $E(X)$,

c $Var(X)$,

d the cumulative distribution function of X, for all x,

e $P(3.5 < X < 5.5)$,

f $P(X = 4)$.

2 The continuous random variable X has p.d.f. as shown in the diagram.

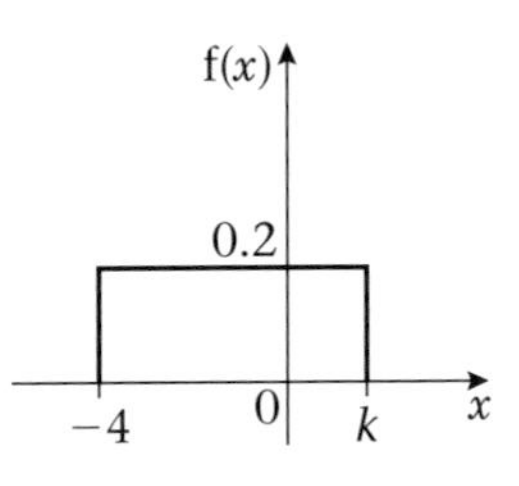

Find

a the value of k,

b $P(-2 < X < -1)$,

c $E(X)$,

d $Var(X)$,

e the cumulative distribution function of X, for all x.

3 The continuous random variable Y is uniformly distributed on the interval $a \leqslant Y \leqslant b$.
Given $E(Y) = 2$ and $Var(Y) = 3$.
Find

a the value of a and the value of b,

b $P(X > 1.8)$.

4 A child has a pair of scissors and a piece of string 20 cm long, which has a mark on one end. The child cuts the string, at a randomly chosen point, into two pieces. Let X represent the length of the piece of string with the mark on it.

a Write down the name of the probability distribution of X and sketch the graph of its probability density function.

b Find the values of $\mathrm{E}(X)$ and $\mathrm{Var}(X)$.

c Using your model, calculate the probability that the shorter piece of string is at least 8 cm long.

5 Joan records the temperature every day. The highest temperature she recorded was 29 °C to the nearest degree. Let X represent the error in the measured temperature.

a Suggest a suitable model for the distribution of X.

b Using your model calculate the probability that the error will be less than 0.2 °C.

c Find the variance of the error in the measured temperature.

6 Jameil catches a bus to work every morning. According to the timetable the bus is due at 9 a.m., but Jameil knows that the bus can arrive at a random time between three minutes early and ten minutes late. The random variable X represents the time, in minutes, after 9 a.m. when the bus arrives.

a Suggest a suitable model for the distribution of X and specify it fully.

b Calculate the mean value of X.

c Find the cumulative distribution function of X.

Jameil will be late for work if the bus arrives after 9.05 a.m.

d Find the probability that Jameil is late for work.

7 A plumber measures, to the nearest cm, the lengths of pipes.

a Suggest a suitable model to represent the difference between the true lengths and the measured lengths.

b Find the probability that for a randomly chosen rod the measured length will be within 0.2 cm of the true length.

c Three pipes are selected at random. Find the probability that all three pipes will be within 0.2 cm of the true length.

8 A coffee machine dispenses coffee into cups. It is electronically controlled to cut off the flow of coffee randomly between 190 ml and 210 ml. The random variable X is the volume of coffee dispensed into a cup.

a Specify the probability density function of X and sketch its graph.

b Find the probability that the machine dispenses

i less than 198 ml,

ii exactly 198 ml.

c Calculate the inter-quartile range of X.

9 Write down the name of the distribution you would recommend as a suitable model for each of the following situations.

a the difference between the true height and the height measured, to the nearest cm, of randomly chosen people.

b the heights of randomly selected 18-year-old females.

Summary of key points

1 A random variable having a continuous uniform distribution over the interval (a, b) has p.d.f.

$$f(x) = \begin{cases} \dfrac{1}{b-a}, & a < x < b, \\ 0, & \text{otherwise.} \end{cases}$$

2 For a random variable X having a uniform distribution

$$E(X) = \frac{a+b}{2}$$

$$\text{Var}(X) = \frac{(b-a)^2}{12}$$

$$F(x) = \begin{cases} 0, & x < a, \\ \dfrac{x-a}{b-a}, & a \leqslant x \leqslant b, \\ 1, & x > b. \end{cases}$$

After studying this chapter you should

- know when and how to apply a continuity correction
- know how to approximate a binomial distribution with a normal distribution
- know how to approximate a Poisson distribution with a normal distribution.

Sometimes the calculations using a binomial or Poisson distribution can become quite cumbersome because of the numbers involved. In many cases the normal distribution provides a simple and accurate approximation of the probability required.

5 Normal approximations

This chapter will show you how to find out how often a particular number of boats are hired from a marina in a given period.

5.1 Using a continuity correction.

The binomial and Poisson distributions are both discrete distributions but the Normal distribution is continuous. This means that, for a discrete distribution, $P(X \leqslant 5)$ and $P(X < 5)$ give different answers as $P(X < 5) = P(X \leqslant 4)$. However, for a continuous distribution such as the normal, $P(X \leqslant 5)$ and $P(X < 5)$ would have the same value since $P(X = 5)$ is zero.

If you are approximating a discrete distribution, X, by a continuous distribution, Y, you need to consider how to treat the decimal values of Y between the discrete values of X.

From GCSE, you will be familiar with the idea that, if $X = 5$ to the nearest integer, then $4.5 \leqslant X < 5.5$ and we use this idea when approximating discrete distributions by a normal (or any continuous) distribution and we call it a **continuity correction**.

Example 1

If X is a discrete distribution, apply a continuity correction to the following probabilities.
a $P(X \leqslant 6)$, **b** $P(X > 10)$, **c** $P(2 \leqslant X < 5)$.

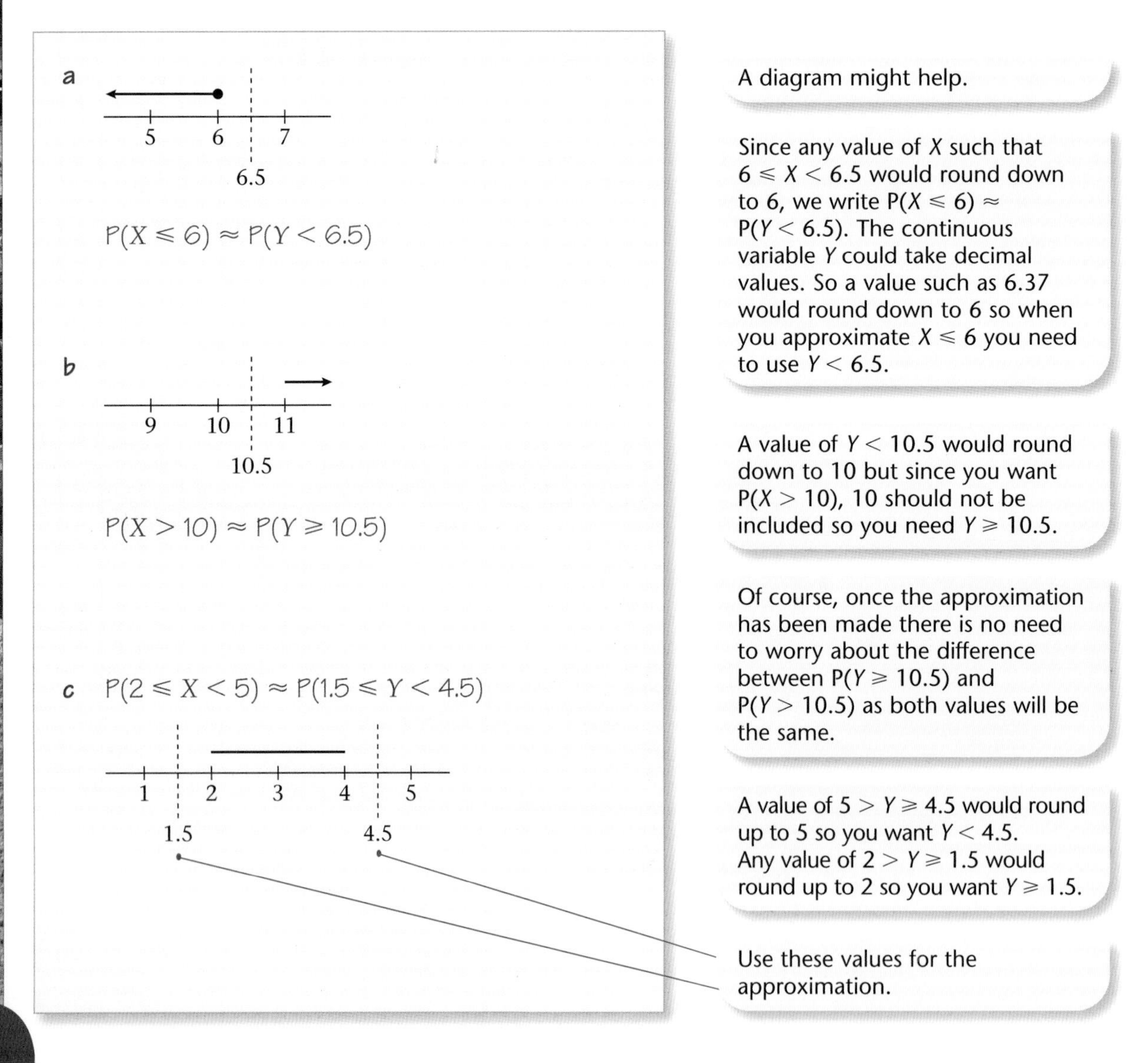

The continuity correction can be summarised in the following simple rules

1 First write your probability using $\leqslant$ or $\geqslant$.
2 For $P(X \leqslant n)$ you simply approximate by $P(Y < n + 0.5)$.
For $P(X \geqslant n)$ you simply approximate by $P(Y \geqslant n - 0.5)$.

You may find these rules helpful but the required continuity correction can always be found using a simple diagram.

Exercise 5A

1 The discrete random variable X takes integer values and is to be approximated by a normal distribution. Apply a continuity correction to the following probabilities.

a $P(X \leqslant 7)$ **b** $P(X < 10)$ **c** $P(X > 5)$
d $P(X \geqslant 3)$ **e** $P(17 \leqslant X \leqslant 20)$ **f** $P(18 < X < 30)$
g $P(28 < X \leqslant 40)$ **h** $P(23 \leqslant X < 35)$

5.2 Approximating a binomial distribution by a normal distribution.

The random variable $X \sim B(n, p)$.

If n is large, calculating binomial coefficients can be difficult and if n is very large most calculators will be unable to perform the calculation (try ${}^{500}C_{240}$ on your calculator).

If p is close to 0.5, then the binomial distribution will be fairly symmetric and so it is reasonable to assume that a normal distribution might provide a suitable approximation.

It is clearly sensible to choose a normal distribution that has the same mean and variance as the original binomial distribution and this leads to the following approximation.

If $X \sim B(n, p)$ and

- **n is large**
- **p is close to 0.5**
Then X can be approximated by $Y \sim N(np, np(1 - p))$
i.e. $\sigma = \sqrt{np(1 - p)}$

The mean and variance of a binomial were given in Section 1.5.

There is no definitive answer to the question of how large n should be or how close to 0.5 p should be. The simple answer is that the larger n is and the closer p is to 0.5 the better. However, the approximation does work quite well for relatively small values of n if p is close to 0.5, as the following example illustrates.

Example 2

The random variable $X \sim B(20, 0.4)$.

a Use tables to find $P(X \leqslant 6)$.

b Use a normal approximation to estimate $P(X \leqslant 6)$.

a $P(X \leqslant 6) = 0.2500$

Use the cumulative binomial distribution tables with $n = 20$, $p = 0.4$ and $x = 0.6$.

b $X \sim B(20, 0.4)$ so $Y \sim N(8, (\sqrt{4.8})^2)$

Using $\mu = np = 20 \times 0.4 = 8$ and $\sigma^2 = np(1 - p) = 20 \times 0.4 \times 0.6 = 4.8$. So $\sigma = \sqrt{4.8}$.

$P(X \leqslant 6) \approx P(Y < 6.5)$

Apply the continuity correction.

$P(Y < 6.5) = P\left(Z < \frac{6.5 - 8}{\sqrt{4.8}}\right)$

Standardise using $z = \frac{x - \mu}{\sigma}$.

$= P(Z < -0.6846\ldots)$

Use 0.68 in the tables as this is the nearest value.

$= 1 - 0.7517$

$= 0.2483$

Remember the normal tables are in the S1 section of the formula booklet.

NB A calculator would give 0.24678... here but both of these values are equal to 0.25 to 2 s.f. thus demonstrating that the approximation has worked quite well.

Example 3

The random variable $X \sim B(120, 0.25)$. Use a normal approximation to estimate $P(35 \leqslant X \leqslant 45)$.

$X \sim B(120, 0.25)$ so $Y \sim N(30, (\sqrt{22.5})^2)$

$\mu = 120 \times 0.25 = 30$
$\sigma^2 = 120 \times 0.25 \times 0.75 = 22.5$

$P(35 \leqslant X \leqslant 45) \approx P(34.5 \leqslant Y \leqslant 45.5)$

Apply continuity corrections.

$= P\left(\frac{34.5 - 30}{\sqrt{22.5}} \leqslant Z < \frac{45.5 - 30}{\sqrt{22.5}}\right)$

Standardise using $z = \frac{x - \mu}{\sigma}$.

$= P(0.9486\ldots \leqslant Z < 3.27\ldots)$

Use 0.95 and 3.25 as these are the nearest values in the tables.

$= 0.9994 - 0.8289$

$= 0.1705$

NB Calculators will give 0.17084... here so the examiners would usually accept answers which round to 0.171.

Some calculators will give an answer to the original question using a binomial distribution (0.1700 in this case). However, if a question tells you to use a normal (or a suitable) approximation then simply giving this answer will usually score no marks. This calculator value can provide a useful check though and that is sometimes reassuring in an examination.

Example 4

For a particular type of flower bulb, 70% will produce yellow flowers. A random sample of 80 bulbs is planted.

Calculate the percentage error incurred when using a normal approximation to estimate the probability that there are exactly 50 yellow flowers.

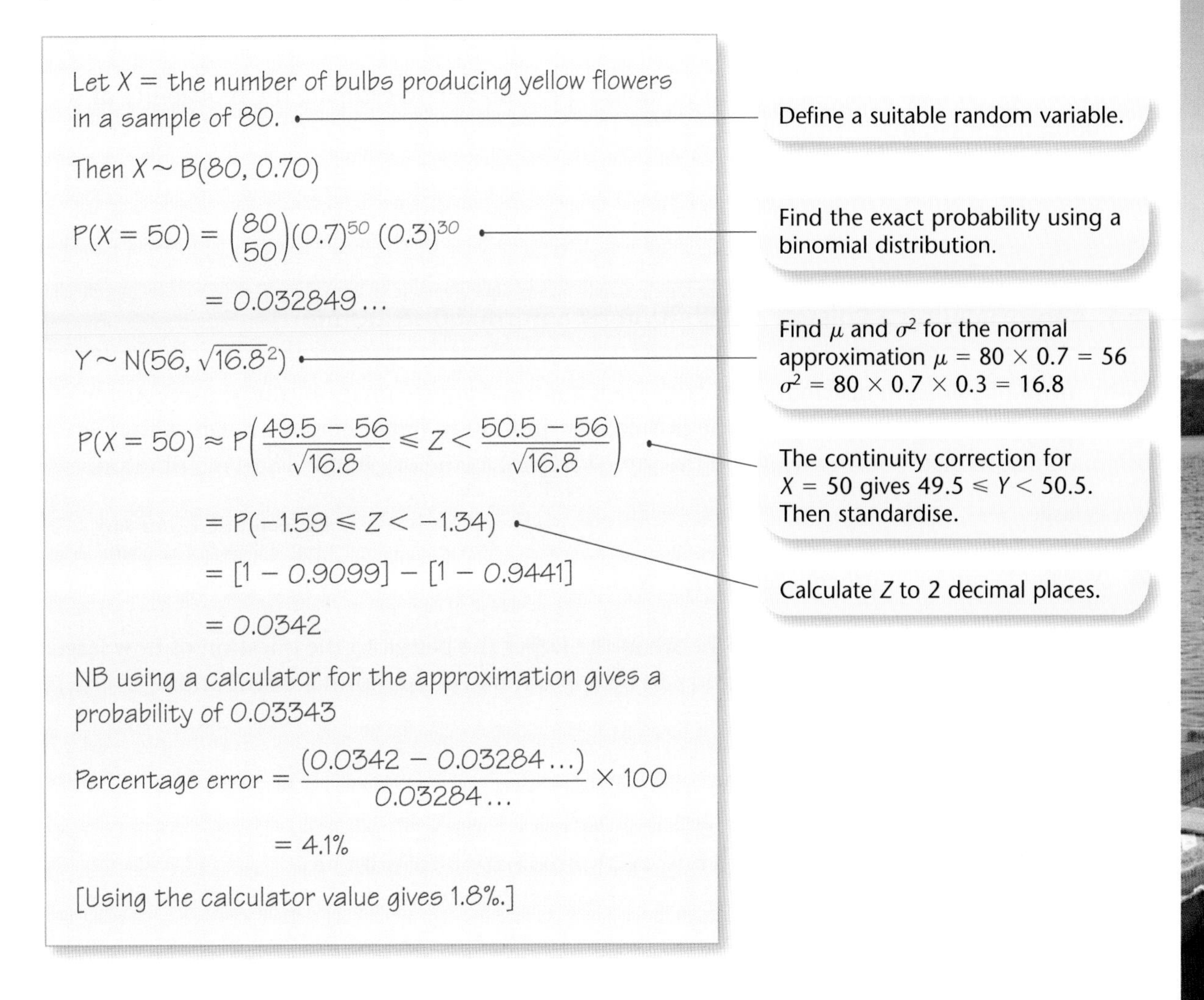

Let X = the number of bulbs producing yellow flowers in a sample of 80.

Define a suitable random variable.

Then $X \sim B(80, 0.70)$

$$P(X = 50) = \binom{80}{50}(0.7)^{50}(0.3)^{30}$$

Find the exact probability using a binomial distribution.

$$= 0.032849\ldots$$

$$Y \sim N(56, \sqrt{16.8}^2)$$

Find μ and σ^2 for the normal approximation $\mu = 80 \times 0.7 = 56$
$\sigma^2 = 80 \times 0.7 \times 0.3 = 16.8$

$$P(X = 50) \approx P\left(\frac{49.5 - 56}{\sqrt{16.8}} \leqslant Z < \frac{50.5 - 56}{\sqrt{16.8}}\right)$$

The continuity correction for $X = 50$ gives $49.5 \leqslant Y < 50.5$. Then standardise.

$$= P(-1.59 \leqslant Z < -1.34)$$

Calculate Z to 2 decimal places.

$$= [1 - 0.9099] - [1 - 0.9441]$$

$$= 0.0342$$

NB using a calculator for the approximation gives a probability of 0.03343

$$\text{Percentage error} = \frac{(0.0342 - 0.03284\ldots)}{0.03284\ldots} \times 100$$

$$= 4.1\%$$

[Using the calculator value gives 1.8%.]

Exercise 5B

1 The random variable $X \sim B(150, \frac{1}{3})$. Use a suitable approximation to estimate

a $P(X \leqslant 40)$, **b** $P(X > 60)$, **c** $P(45 \leqslant X \leqslant 60)$.

2 The random variable $X \sim B(200, 0.2)$. Use a suitable approximation to estimate

a $P(X < 45)$, **b** $P(25 \leqslant X < 35)$, **c** $P(X = 42)$.

3 The random variable $X \sim B(100, 0.65)$. Use a suitable approximation to estimate

a $P(X > 58)$, **b** $P(60 < X \leqslant 72)$, **c** $P(X = 70)$.

4 Sarah rolls a fair die 90 times. Use a suitable approximation to estimate the probability that the number of sixes she obtains is over 20.

5 In a multiple choice test there are 4 possible answers to each question. Given that there are 60 questions on the paper, use a suitable approximation to estimate the probability of getting more than 20 questions correct if the answer to each question is chosen at random from the 4 available choices for each question.

6 A fair coin is tossed 70 times. Use a suitable approximation to estimate the probability of obtaining more than 45 heads.

5.3 Approximating a Poisson distribution by a normal distribution.

If the mean of a Poisson distribution is large then a normal approximation can be used. As for the binomial distribution we choose the normal distribution to have the same mean as the original Poisson distribution mean and the same variance as the original Poisson distribution.

If λ is large

■ **$X \sim$ Po(λ) can be approximated by $Y \sim$ N(λ, $(\sqrt{\lambda})^2$)**

In Section 2.2 you saw that if $X \sim \text{Po}(\lambda)$ then $\text{E}(X) = \lambda$ and $\text{Var}(X) = \lambda$.

As before there is no simple answer, other than the larger the better, to the question of how large λ should be before a Poisson distribution can be approximated to a normal distribution.

Example 5

The random variable $X \sim \text{Po}(25)$. Use a normal approximation to estimate

a $\text{P}(X > 30)$, **b** $\text{P}(18 \leqslant X < 35)$.

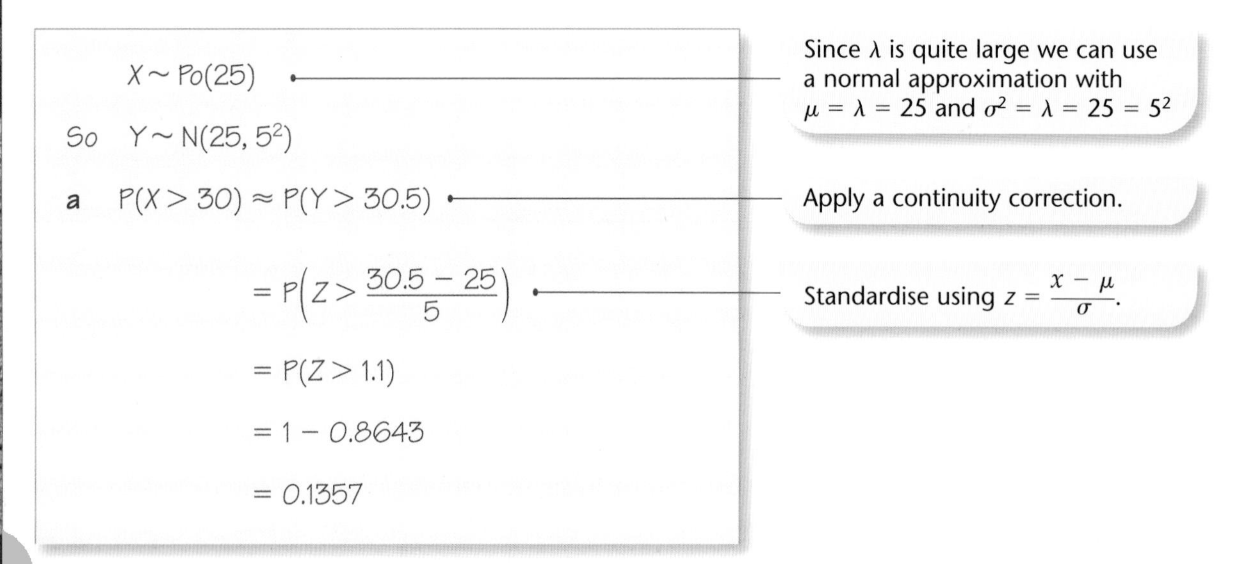

$X \sim \text{Po}(25)$

So $Y \sim \text{N}(25, 5^2)$

Since λ is quite large we can use a normal approximation with $\mu = \lambda = 25$ and $\sigma^2 = \lambda = 25 = 5^2$

a $\text{P}(X > 30) \approx \text{P}(Y > 30.5)$

Apply a continuity correction.

$= \text{P}\left(Z > \frac{30.5 - 25}{5}\right)$

Standardise using $z = \frac{x - \mu}{\sigma}$.

$= \text{P}(Z > 1.1)$

$= 1 - 0.8643$

$= 0.1357$

b $P(18 \leqslant X < 35) \approx P(17.5 \leqslant Y < 34.5)$ — Apply continuity correction and then standardise.

$$= P\left(\frac{17.5 - 25}{5} \leqslant Z < \frac{34.5 - 25}{5}\right)$$

$$= P(-1.5 \leqslant Z < 1.9)$$

$$= 0.4713 + 0.4332$$

$$= 0.9045$$

NB Your calculator may be able to calculate these probabilities using the original Poisson distribution (part **a** would be 0.13669… and **b** 0.90568) but, in S2, if you are asked to use an approximation you will not receive any credit for simply giving these values.

Example 6

A car hire firm has a large fleet of cars for hire by the day and it is found that the fleet suffers breakdowns at the rate of 21 per week. Assuming that breakdowns occur at a constant rate, randomly in time and independently of one another, use a suitable approximation to estimate the probability that in any one week more than 27 breakdowns occur.

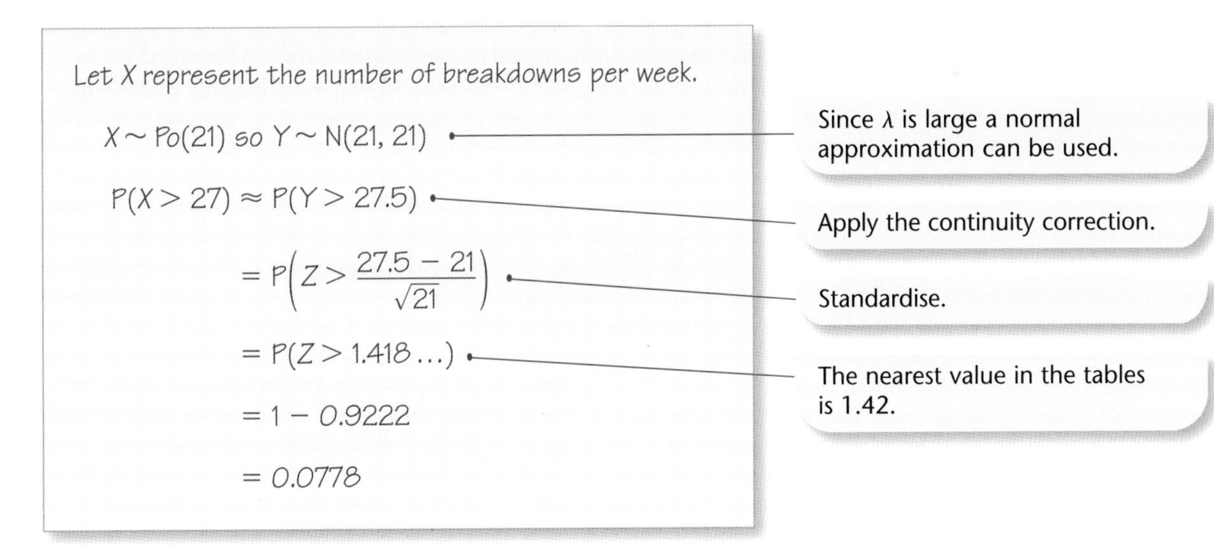

Let X represent the number of breakdowns per week.

$X \sim \text{Po}(21)$ so $Y \sim \text{N}(21, 21)$ — Since λ is large a normal approximation can be used.

$P(X > 27) \approx P(Y > 27.5)$ — Apply the continuity correction.

$$= P\left(Z > \frac{27.5 - 21}{\sqrt{21}}\right)$$ — Standardise.

$$= P(Z > 1.418\ldots)$$ — The nearest value in the tables is 1.42.

$$= 1 - 0.9222$$

$$= 0.0778$$

NB A calculator would give 0.078034…so the examiners would probably accept answers which round to 0.078.

Exercise 5C

1 The random variable $X \sim \text{Po}(30)$. Use a suitable approximation to estimate

a $P(X \leqslant 20)$, **b** $P(X > 43)$, **c** $P(25 \leqslant X \leqslant 35)$.

2 The random variable $X \sim \text{Po}(45)$. Use a suitable approximation to estimate

a $\text{P}(X < 40)$, **b** $\text{P}(X \geqslant 50)$, **c** $\text{P}(43 < X \leqslant 52)$.

3 The random variable $X \sim \text{Po}(60)$. Use a suitable approximation to estimate

a $\text{P}(X \leqslant 62)$, **b** $\text{P}(X = 63)$, **c** $\text{P}(55 \leqslant X < 65)$.

4 The disintegration of a radioactive specimen is known to be at the rate of 14 counts per second. Using a normal approximation for a Poisson distribution, determine the probability that in any given second the counts will be

a 20, 21 or 22, **b** greater than 10, **c** above 12 but less than 16.

5 A marina hires out boats on a daily basis. The mean number of boats hired per day is 15. Using the normal approximation for a Poisson distribution, find, for a period of 100 days

a how often 5 or fewer boats are hired,

b how often exactly 10 boats are hired,

c on how many days they will have to turn customers away if the marina owns 20 boats.

5.4 Choosing the appropriate approximation.

In Section 2.5 you saw that a binomial distribution can sometimes be approximated by a Poisson distribution and in Section 5.2 you saw that sometimes it can be approximated by a normal distribution. The approximations you need in S2 can be summarised by the following diagram.

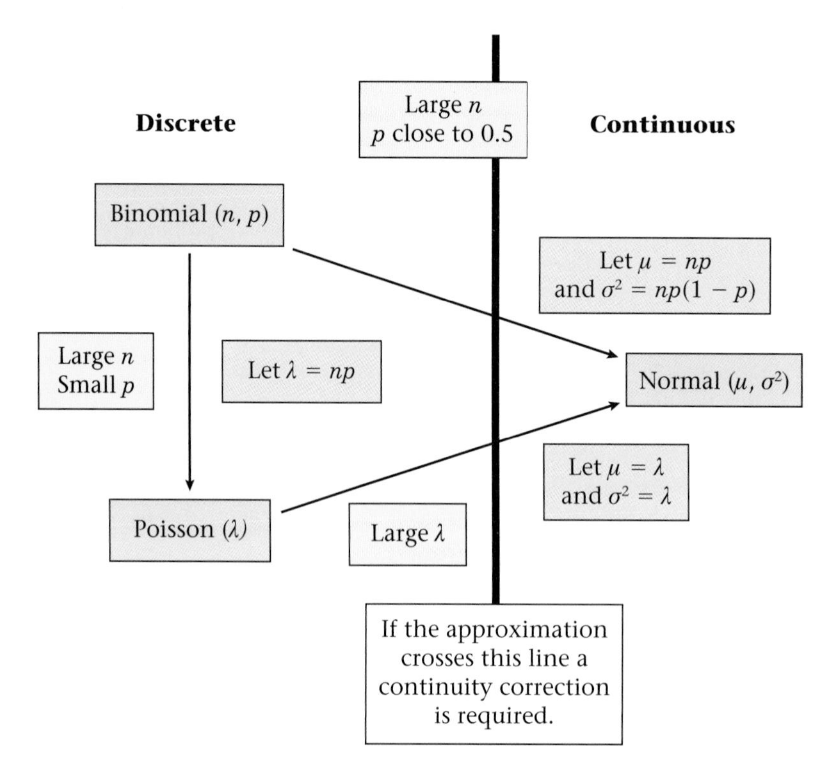

If you are approximating a binomial distribution to a normal distribution you should always go directly to the normal distribution and not via a Poisson distribution as this involves one not two approximations and should therefore be more accurate.

So for a binomial distribution there are two possible approximations, depending upon whether p lies close to 0.5 (in which case a normal distribution is used) or p is small (in which case a Poisson distribution is used). If you are in doubt over which approximation is appropriate a useful 'rule of thumb' is to calculate the mean np and, if this is less than or equal to 10, you should be able to use the Poisson tables so that approximation can be used. If the mean is more than 10 then a normal approximation is usually suitable.

Example 7

A spinner is designed to land on red 10% of the time. Use suitable approximations to estimate the probability of

a fewer than 4 reds in 60 turns of the spinner,

b more than 20 reds in 150 turns of the spinner.

a Let X = the number of reds in 60 turns of the spinner.

$X \sim B(60, 0.1)$

Calculate the mean of the binomial.

$E(X) = 60 \times 0.1 = 6$ so a Poisson approximation can be used.

So $X \approx \sim Po(6)$

No continuity correction is required – simply use the Poisson tables with $\lambda = 6$ and $x = 3$.

$P(X < 4) = P(X \leqslant 3)$

$= 0.1512$

b Let R = the number of reds in 150 turns of the spinner.

$R \sim B(150, 0.1)$

$E(R) = 150 \times 0.1 = 15$ so use a normal approximation.

Since the mean is >10 use a normal approximation. The value of $\sigma^2 = 150 \times 0.1 \times 0.9 = 13.5$ and a continuity correction is needed.

So $X \approx \sim N(15, (\sqrt{13.5})^2)$

$P(R > 20) \approx P(Y \geqslant 20.5)$

$= P\left(Z > \dfrac{20.5 - 15}{\sqrt{13.5}}\right)$

$= P(Z > 1.496\ldots)$

Use $z = 1.50$ as this is the nearest value in the tables.

$= 1 - 0.9332$

$= 0.0668$

NB The exact binomial probabilities in these cases are **a** 0.13739… and **b** 0.072088.

Mixed exercise 5D

1 A fair die is rolled and the number of sixes obtained is recorded.
Using suitable approximations, find the probability of

a no more than 10 sixes in 48 rolls of the die,

b at least 25 sixes in 120 rolls of the die.

2 A fair coin is spun 60 times.
Use a suitable approximation to estimate the probability of obtaining fewer than 25 heads.

3 The owner of a local corner shop calculates that the probability of a customer buying a newspaper is 0.40 but the proportion of customers who spend over £10 is 0.04.
A random sample of 100 customers' shopping is recorded. Use suitable approximations to estimate the probability that in this sample

a at least half of the customers bought a newspaper,

b more than 5 of them spent over £10.

4 Street light failures in a town occur at a rate of one every two days. Assuming that X, the number of street light failures per week, has a Poisson distribution, find the probabilities that the number of street lights that will fail in a given week is

a exactly 2,

b less than 6.

Using a suitable approximation estimate the probability that

c there will be fewer than 45 street light failures in a 10-week period.

5 Past records from a supermarket show that 20% of people who buy chocolate bars buy the family size bar. A random sample of 80 people is taken from those who had bought chocolate bars.

a Use a suitable approximation to estimate the probability that more than 20 of these 80 bought family size bars.

The probability of a customer buying a gigantic chocolate bar is 0.02.

b Using a suitable approximation estimate the probability that fewer than 5 customers in a sample of 150 buy a gigantic chocolate bar.

Summary of key points

1 The random variable $X \sim B(n, p)$ can be approximated by $Y \sim N(np, np(1-p))$ when

n is large
p is close to 0.5

A continuity correction should be used.

2 The random variable $X \sim Po(\lambda)$ can be approximated by $Y \sim N(\lambda, \lambda)$ when

λ is large

A continuity correction should be used.

After completing this chapter you should

- know what is meant by a population
- know the difference between a census and a sample
- know what is meant by a sampling frame
- know what is meant by a sampling unit
- know the advantages and disadvantages of taking a census and taking a sample
- know what is meant by the sampling distribution of a statistic
- be able to find the sampling distribution of samples in simple cases.

6 Populations and samples

Consider articles produced by a machine. Some may be faulty and some may be satisfactory.
By taking a number of articles (a sample) and finding how many of them are faulty we can draw conclusions about the number of faulty articles among all of the articles being produced (the population).

6.1 Populations, censuses and samples.

■ A population is a collection of individual people or items. Examples include manufactured items, plants, supermarkets, members of a club, etc.

■ A population may be of finite size. For example, the number of students at a particular school is finite because you can give each student a number and count how many of them there are.

■ A population may be considered to be of infinite size if it is impossible to know exactly how many members there are in the population. For example the number of grains of sand on a stretch of beach. It would be impossible to count them all.

■ A population may be of countably infinite size if we know that the population could be infinite but in practice we can count the number of individual members of it. For example the number of throws of a dice in order to obtain a 6 is countably infinite.

■ If information is to be obtained from all members of the population, the investigation is known as a census.

Example 1

Give an example of the following types of population

a finite **b** infinite.

a The number of students in a sixth form in the academic year 2007/2008 is a finite population.

This is a small population. By contrast the yearly output of a machine would be a large but finite population.

b The number of telephone calls made in a year throughout the world can be considered infinite. (It would be impossible to count them.)

These are not the only possible answers. Perhaps you can think up some for yourself.

6.2 The advantages and disadvantages of taking a census.

■ The advantages of taking a census are
- every single member of the population is used
- it is unbiased
- it gives an accurate answer.

■ The disadvantages of taking a census are
- it takes a long time to do
- it is costly
- it is often difficult to ensure that the whole population is surveyed.

6.3 Understanding the concept of sampling.

For a large population where a census is ruled out as being impractical the required information is obtained from selected members that form a sub-set of the population.

The chosen members are referred to as a **sample**.

An investigation using a sample is called a **sample survey**.

The individual units of a population are known as **sampling units**.

For example, John, Sandra and Tony could be three of the sampling units making up the population of students in a sixth form.

When the sampling units within a population are individually named or numbered to form a list then this list of sampling units is called a **sampling frame**.

A sampling frame can take a variety of forms – a list, index, map, file, database – but whatever its form, how well a sampling frame covers a population and its accuracy are important as the sampling frame is the basis of any sample drawn.

For example, the class register of all the students in the sixth form would be a sampling frame.

Example 2

a Give two reasons why a census may not be used.

A factory manager is suggesting that overalls with the company logo be provided for workers. It is decided to ask a sample of workers for their opinion about the proposed change.

b Suggest what would be suitable sampling units for this investigation.

c Suggest a suitable sampling frame.

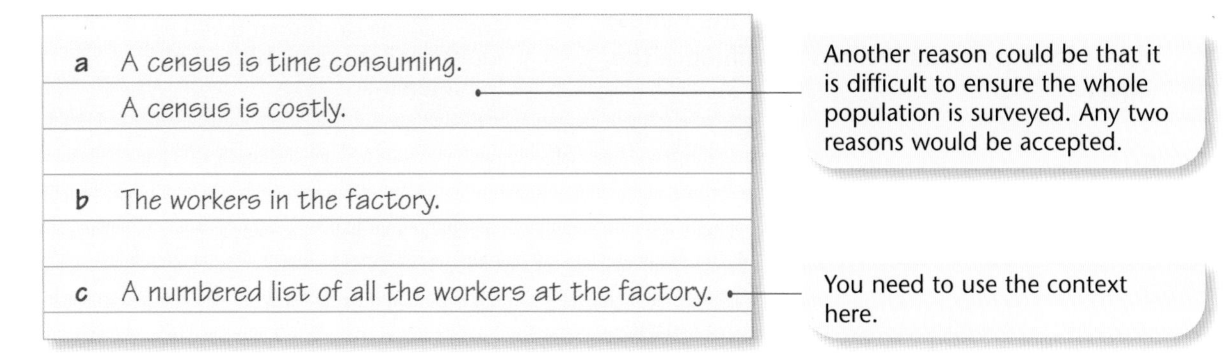

a A census is time consuming.
A census is costly.

Another reason could be that it is difficult to ensure the whole population is surveyed. Any two reasons would be accepted.

b The workers in the factory.

c A numbered list of all the workers at the factory.

You need to use the context here.

6.4 The advantages and disadvantages of sampling.

The advantages of sampling

- **If a population is large and well mixed a sample will be *representative* of the whole population. This has many advantages.**

For example, if a sample is taken from a production line producing breakfast cereal, the fat content of the sample is representative of the fat content of all the cereal being produced.

- **Sampling is generally cheaper than taking a census.**

Checking every item on a production line could make the process uneconomic.

- **Sampling is advantageous where the testing of items results in their destruction.**

For example, testing the life-time of light bulbs.

- **When using a sample rather than a census, data is generally more readily available.**

The disadvantages of sampling

- **There is an uncertainty about sampling in that there will be a natural variation between any two samples due to the *natural variation* between the individual units that make up the samples.**

For example, if you picked out three individuals from your class, and then picked out another three, you just might get the three smallest people in the first sample and the three tallest in the second sample.

- **Another source of uncertainty is called bias. Bias is anything which occurs when taking a sample which prevents it from being truly representative of the population from which it is taken.**

Bias can occur if you sample from an incomplete sampling frame, or if you get responses only from people that have a particular interest in the topic being studied.

For example, a sample chosen from the people getting off a train cannot provide a good representation of the views of people who drive into the city.

Bias can occur if the person taking the sample allows their personal feelings to influence the choice.

For example, a student asked to interview people might only pick people in his family.

Exercise 6A

1 **a** Write down a brief description of a census.

b Write down two advantages of using a census rather than a sample.

c Write down two disadvantages of using a census rather than a sample.

2 Write down which of the following are finite populations and which are infinite populations.

a Stars in the sky.

b Workers in a supermarket.

c The number of cows in Farmer Jacob's herd of cows.

3 **a** Write down a brief description of a sample.

b Write down one disadvantage of taking a sample rather than a census.

c Write down two advantages of taking a sample rather than a census.

4 A city council wants to know what people think about its recycling centre. The council decides to carry out a sample survey to get the opinion of resident's views.

a Write down one reason why the council should not take a census.

b Suggest a suitable sampling frame.

c Identify the sampling units.

5 A factory manufactures climbing ropes. The manager of the factory decides to investigate the breaking point of the ropes.

Write down a reason, other than easier and cheaper, why he would not use a census.

6 A supermarket manager wants to find out whether customers are satisfied with the range of products in the supermarket. He decides to do a survey.

a Write down a reason why the manager decides to use a sample rather than a census.

He decides to do a sample survey.

b Describe the sampling units for the sample survey.

c Give one advantage and one disadvantage of using a sample survey.

7 A manager of a garage wants to know what his mechanics think about a new pension scheme designed for them. He decides to ask all the mechanics in the garage.

a Describe the population he will use.

b Write down the main advantage there will be in asking all his mechanics.

8 Each computer produced by a manufacturer is stamped with a unique serial number. ITPro Limited make their computers in batches of 1000. Before selling the computers, they test a random sample of 5 to see what electrical overload they will take before breaking down.

a Give one reason, other than to save time and cost, why a sample is taken rather than a census.

b Suggest a suitable sampling frame.

c Identify the sampling units.

6.5 Simple random sampling.

A sample should be a true (unbiased) representation of the population from which it is taken. One way of achieving this is by obtaining the data using a process known as **simple random sampling**.

■ **A simple random sample, of size *n*, is one taken so that every possible sample of size *n* has an equal chance of being selected.**

A well known example of simple random sampling is the National Lottery. A sample of size 6 is taken from the numbers 1 to 49. Every sample has the same chance of winning the jackpot and so each individual lottery ticket has the same chance (of almost 1 in 14 million) of winning. Raffle draws operate in a similar way. Each ticket is placed in a box, the box is shaken to mix the tickets up and tickets are then drawn to determine the winner.

The theory in this and the next chapter is based on the assumption that the population is numerical and that **simple random sampling** is used.

In book S1 you learnt that a random variable represents the outcome obtained as the result of an experiment. A random variable always has a numerical value.

Choosing a number from a population is an experiment (being selected) and the outcome is a number so we can define the outcome of the first member of the sample as being the random variable X_1.

In the same way the outcome of the second and third members of the sample are the random variables X_2 and X_3.

If there are to be n in the sample then the nth member of the sample will be the random variable X_n.

The distribution of a discrete random variable is defined by listing its possible values and the probability with which they occur.

■ **A simple random sample of size *n* consists of the observations $X_1, X_2, \ldots, X_n$ from a population where the X_i**
- **are independent random variables**
- **have the same distribution as the population.**

Following previous custom we will use x_1, x_2, x_3 etc. to describe particular values of the random variables X_1, X_2, X_3, etc.

For example the population might consist of a set of 48 balls in a bag numbered from 1 to 48. X_1 can take any of the values between 1 and 48, and the probability of getting any number is $\frac{1}{48}$. We can define X_1 as being a uniform distribution over the set of values 1 to 48 with $P(X_1 = x_1) = \frac{1}{48}$.

X_2 and X_3 are also uniform distributions over the set of values 1 to 48 with $P(X_2 = x_2) = \frac{1}{48}$ and $P(X_3 = x_3) = \frac{1}{48}$.

6.6 The concept of a statistic.

Any characteristic of a population which is measurable is called a **population parameter**. (We usually use Greek letters for population parameters.) A parameter is a numerical property of a sample.

For example the population mean, μ, and the population variance, σ^2, are population parameters.

Usually the population is too large to calculate these parameters. In order to estimate a population parameter we take a random sample from the population and use observations from the items in it to estimate the required parameters. For example, we could use the sample mean as an estimate for the population mean.

If we drew the numbers 3, 8, 32, 38, 43 and 44 from our bag the sample mean $\bar{x} = \frac{\Sigma X_i}{n} = \frac{3 + 8 + 32 + 38 + 43 + 44}{6} = 28$ could be an estimate for the parameter μ.

- **A statistic is a quantity calculated solely from the observations in a sample. It does not involve any unknown parameters i.e. a statistic is a numerical property of a sample.**

The sample mean $\bar{X} = \frac{\Sigma X_i}{n}$ and sample variance $S^2 = \frac{\Sigma(X_i - \bar{X})^2}{n - 1}$ are statistics

but $\frac{\Sigma(X_i - \mu)^2}{n - 1}$ is not (it involves the population parameter μ).

Example 3

a Define a statistic.

A random sample $X_1, X_2, \ldots, X_n$ is taken from a population with unknown mean μ.

b For each of the following state whether or not it is a statistic.

i $\frac{X_2 + X_5 + X_8}{3}$ **ii** $\frac{\Sigma X^2}{n} - \left(\frac{\Sigma X}{n}\right)^2$ **iii** $\frac{\Sigma X}{n} - \mu^2$

a A quantity calculated solely from the observations in a sample.

b **i** and **ii** are statistics but **iii** is not (it involves μ which is unknown).

6.7 The sampling distribution of a statistic.

If we repeatedly take samples from a population and calculate the same statistic each time there is a range of values that the statistic can take. The statistic will have its own distribution which we call the **sampling distribution**.

- **The sampling distribution of a statistic gives all the values of a statistic and the probability that each would happen by chance alone.**

Example 4

A school wishes to introduce a school uniform and is seeking to find out the support this idea has among the students at the school. The random variable X is defined as

$$X = \begin{cases} 1, & \text{if the student would support the idea,} \\ 0, & \text{otherwise.} \end{cases}$$

a Suggest a suitable population and the parameter of interest.

A random sample of 15 students is asked if they would support the idea. The random sample is represented by $X_1, X_2, ..., X_{15}$.

b Write down the sampling distribution of the Statistic $Y = \sum_{i=1}^{15} X_i$.

a The population is the responses of all the school students.
In terms of the random variable X, it will consist of 1's or 0's.
The parameter of interest is p, the proportion of the population who support the idea.

b ΣY = the number of students in the sample who support the idea.
Since the sample is random each observation is independent, p is constant, and the responses will be either success(1) or failure(0)
These are the conditions for a binomial distribution. $Y \sim B(15, p)$

See Chapter 1.

Example 5

A manufacturer of light bulbs sells 60 watt and 100 watt bulbs in the ratio of 3 : 1.

a Find the mean and variance of the wattage of the light bulbs in this population.

A random sample of 3 light bulbs is taken from a store containing bulbs in this ratio.

b List all the possible samples.

c Find the sampling distribution of the mean, $\overline{X}$.

d Find the sampling distribution of the mode, M.

a The distribution of the population is

X:	60	100
$P(X = x)$:	$\frac{3}{4}$	$\frac{1}{4}$

$\mu = \Sigma xP(X = x) = 60 \times \frac{3}{4} + 100 \times \frac{1}{4} = 70$ watts

$\sigma^2 = \Sigma x^2P(X = x) - \mu^2 = 60^2 \times \frac{3}{4} + 100^2 \times \frac{1}{4} - (70)^2$

$= 300$ watts^2

b The possible samples are

(60, 60, 60)

(100, 60, 60) (60, 100, 60) (60, 60, 100)

(100, 100, 60) (100, 60, 100) (60, 100, 100)

(100, 100, 100)

Here we have written out all the possible samples of size 3.

$P(X = 60)$ is constant at $\frac{3}{4}$.

c $P(\bar{X} = 60) = (\frac{3}{4})^3 = \frac{27}{64}$

i.e the case (60, 60, 60)

$P(\bar{X} = 73\frac{1}{3}) = 3 \times \frac{1}{4} \times (\frac{3}{4})^2 = \frac{27}{64}$

i.e. cases (100, 60, 60) (60, 100, 60) (60, 60, 100)

$P(\bar{X} = 86\frac{2}{3}) = 3 \times (\frac{1}{4})^2 \times \frac{3}{4} = \frac{9}{64}$

i.e. cases (100, 100, 60) (100, 60, 100) (60, 100, 100)

$P(\bar{X} = 100) = (\frac{1}{4})^3 = \frac{1}{64}$

i.e. the case (100, 100, 100)

The distribution of $\bar{X}$ is

$\bar{X}$:	60	$73\frac{1}{3}$	$86\frac{2}{3}$	100
$P(\bar{x})$:	$\frac{27}{64}$	$\frac{27}{64}$	$\frac{9}{64}$	$\frac{1}{64}$

This shows the distribution of the statistic X by writing down all the possible values and the probability with which each is likely to occur.

d The mode M can only take the values 60 and 100.

$P(M = 60) = \frac{27}{64} + \frac{27}{64} = \frac{27}{32}$

i.e. cases (60, 60, 60) (100, 60, 60) (60, 100, 60) (60, 60, 100)

$P(M = 100) = \frac{9}{64} + \frac{1}{64} = \frac{5}{32}$

i.e. the other cases.

The distribution of M is:

M:	60	100
$p(m)$:	$\frac{27}{32}$	$\frac{5}{32}$

Example 6

A large bag of counters has 10% with the number 0 on, 40% with the number 1 on and 50% with the number 2. A random sample of 3 counters is taken from the bag.

a List all possible samples.

b Find the sampling distribution of the median N.

a The possible samples are

$(0, \underline{0}, 0)$ $(0, \underline{0}, 1)$ $(0, 1, \underline{0})$ $(1, \underline{0}, 0)$ $(0, \underline{0}, 2)$ $(0, 2, \underline{0})$ $(2, \underline{0}, 0)$ — Median = 0 in these

$(1, \underline{1}, 0)$ $(1, 0, \underline{1})$ $(0, \underline{1}, 1)$ $(1, \underline{1}, 2)$ $(1, 2, \underline{1})$ $(2, \underline{1}, 1)$ $(1, \underline{1}, 1)$ — Median = 1 in these

$(0, \underline{1}, 2)$ $(0, 2, \underline{1})$ $(\underline{1}, 0, 2)$ $(\underline{1}, 2, 0)$ $(2, \underline{1}, 0)$ $(2, 0, \underline{1})$ — Median = 1 in these

$(2, \underline{2}, 0)$ $(2, 0, \underline{2})$ $(0, 2, \underline{2})$ $(\underline{2}, 2, 1)$ $(2, 1, \underline{2})$ $(1, \underline{2}, 2)$ $(2, \underline{2}, 2)$ — Median = 2 in these

b The median can only take the values 0, 1 and 2.

Let $P(0) = p = \frac{1}{10}$, $P(1) = q = \frac{4}{10}$ and $P(2) = r = \frac{5}{10}$

$$P(N = 0) = P(0,0,0) + P(0,0,1) + P(0,1,0) + P(1,0,0) + P(0,0,2) + P(0,2,0) + P(2,0,0)$$

$$= ppp + ppq + pqp + qpp + ppr + prp + rpp$$

$$= \left(\frac{1\times1\times1}{1000}\right) + \left(\frac{1\times1\times4}{1000}\right) + \left(\frac{1\times4\times1}{1000}\right) + \left(\frac{4\times1\times1}{1000}\right) + \left(\frac{1\times1\times5}{1000}\right) + \left(\frac{1\times5\times1}{1000}\right) + \left(\frac{5\times1\times1}{1000}\right)$$

$$= \frac{28}{1000} = \frac{7}{250}$$

$$P(N = 1) = P(1,1,0) + P(1,0,1) + P(0,1,1) + P(1,1,2) + P(1,2,1) + P(2,1,1) + P(1,1,1)$$
$$+ P(0,1,2) + P(0,2,1) + P(1,0,2) + P(1,2,0) + P(2,1,0) + P(2,0,1)$$

$$= qqp + ppq + pqq + qqr + qrq + rqq + qqq$$
$$+ pqr + prq + qpr + qrp + rqp + rpq$$

$$= \left(\frac{4\times4\times1}{1000}\right) + \left(\frac{4\times1\times4}{1000}\right) + \left(\frac{1\times4\times4}{1000}\right) + \left(\frac{4\times4\times5}{1000}\right) + \left(\frac{4\times5\times4}{1000}\right) + \left(\frac{5\times4\times4}{1000}\right) + \left(\frac{4\times4\times4}{1000}\right)$$
$$+ \left(\frac{1\times4\times5}{1000}\right) + \left(\frac{1\times5\times4}{1000}\right) + \left(\frac{4\times1\times5}{1000}\right) + \left(\frac{4\times5\times1}{1000}\right) + \left(\frac{5\times4\times1}{1000}\right) + \left(\frac{5\times1\times4}{1000}\right) = \frac{59}{125}$$

$$P(N = 2) = 1 - \left(\frac{7}{250} + \frac{118}{250}\right) = \frac{125}{250} = \frac{1}{2}$$

The distribution of N is

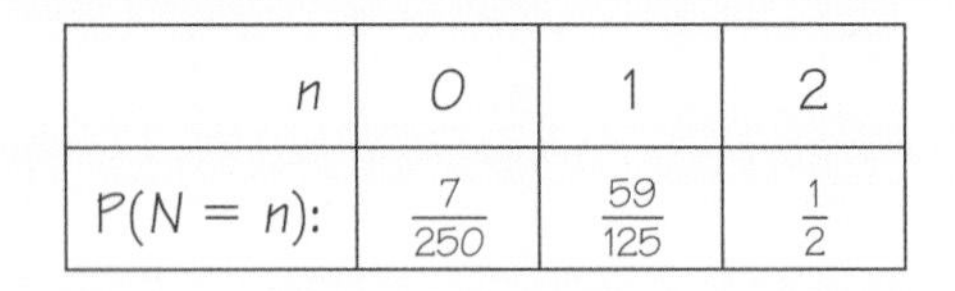

n	0	1	2
$P(N = n)$:	$\frac{7}{250}$	$\frac{59}{125}$	$\frac{1}{2}$

Exercise 6B

1 A forester wants to estimate the height of the trees in a forest. He measures the heights of 50 randomly selected trees and works out the mean height. State with a reason whether or not this mean is a statistic.

2 A random sample $M_1, M_2, M_3, \ldots, M_n$ is taken from a population with unknown mean μ. For each of the following state whether or not it is a statistic.

i $\frac{M_3 + M_8}{2}$ **ii** $\frac{\Sigma M}{n}$ **iii** $\frac{\Sigma M}{n} - \mu^2$

3 The owners of a chain of hairdressing shops want to introduce the use of overalls in all the shops. The random variable Y is defined as

$Y = 0$ if the staff are happy to wear the overalls and

$Y = 1$ if the staff are unhappy about wearing the overalls.

a Suggest a suitable population and identify any parameter of interest.

A random sample of 20 of the hairdressers are asked whether they are happy or unhappy about wearing the overalls.

b Write down the name of the sampling distribution of the statistic $X = \sum_{1}^{20} Y$.

4 A secretary makes spelling mistakes at the rate of 5 for every 10 pages. He has just finished typing a six-page document.

a Write down a suitable sampling distribution for the number of spelling mistakes in his document.

b Find the probability that there has been fewer than 2 spelling mistakes in the document.

5 A bag contains a large number of coins. 50% are 50 pence coins.
25% are 20 pence coins. 25% are 10 pence coins.

a Find the mean, μ, and the variance, σ, for the value of this population of coins.

A random sample of 2 coins is chosen from the bag.

b List all the possible samples that can be chosen.

c Find the sampling distribution for the mean

$$\overline{X} = \frac{X_1 + X_2}{2}.$$

6 A manufacturer makes three sizes of toaster. 40% of the toasters sell for £16, 50% sell for £20 and 10% sell for £30.

a Find the mean and variance of the value of the toasters.

A sample of 2 toasters is sent to a shop.

b List all the possible prices of the samples that could be sent.

c Find the sampling distribution for the mean price $\overline{X}$ of these samples.

7 A supermarket sells a large number of 3-litre and 2-litre cartons of milk. They are sold in the ratio 3:2

a Find the mean and variance of the milk content in this population of cartons.

A random sample of 3 cartons is taken from the shelves (X_1, X_2 and X_3).

b List all the possible samples.

c Find the sampling distribution of the mean $\overline{X}$.

d Find the sampling distribution of the mode M.

e Find the sampling distribution of the median N of these samples.

Mixed exercise 6C

1 A doctor's surgery is to offer health checks to all its patients over 65. In order to estimate the amount of time needed to do these health checks the doctor decides to do the health check for a random sample of 20 patients over 65.

a Write down a suitable sampling frame that the doctor might use.

b Identify the sampling units.

2 The owners of a large gym wish to change the opening hours. They want to find out whether the members will be happy with the new hours. They ask a random sample of 30 members.

a Write two likely reasons why the owners did not ask all the members.

b Suggest a suitable sampling frame.

c Identify the sampling units.

3 **a** Write down a reason why a sampling frame and a population may not be the same.

b Explain briefly why a sample is often used rather than a census.

4 **a** Explain what a statistic is.

A random sample $Y_1, Y_2, \ldots, Y_n$ is taken from a population with unknown mean μ.

b For each of the following state with a reason whether or not it is a statistic.

i $\dfrac{Y_1 + Y_2 + Y_3}{4}$ **ii** $\dfrac{\Sigma Y}{n} - \mu$

5 A company manufactures electric light bulbs. They wish to see how many hours the light bulbs will work before failing. The company decides to test every 200th light bulb coming off the assembly line.

a Write down why the company does not test every light bulb.

b Identify the sampling units.

6 A call centre has 400 people operating the telephones. The manager decides that he needs to know how long the operatives are spending on each call. He times a random sample of 30 operators over one day and works out the mean time per call.

a Write down two advantages of using a sample rather than a census in this case.

b Write down one disadvantage of using a sample in this case.

A sample is to be taken.

c Suggest a sampling frame.

d Identify the sampling units

e Is the mean time the manager works out from the sample a statistic? Give a reason for your answer.

7 A flower shop has ten florists. The owner wants to know whether the florists are happy with the quality of the flowers being delivered to the shop. The owner asks all the florists their views. Write down two reasons why the owner of the florist shop used a census.

8 The weights of tomatoes in a greenhouse are assumed to have mean μ and standard deviation σ.

A sample of 20 tomatoes were each weighed and their weights were recorded. If the sample is represented by $X_1, X_2, \ldots, X_{20}$ state whether or not the following are statistics.

a $\frac{X_1 + X_{20}}{3}$ **b** $\frac{\Sigma X}{20}$ **c** $\Sigma X^2 + \mu$ **d** $\frac{\Sigma X^2}{20} - \sigma^2$

9 A large box of coins contains 5p, 10p, and 20p coins in the ratio 3:2:1.

a Find the mean μ and the variance σ^2 of the value of the coins.

A random sample of 2 coins is taken from the box and their values Y_1 and Y_2 are recorded.

b List all the possible samples that can be taken.

c Find the sampling distribution for the mean $(\bar{Y})$.

10 A bag contains a large number of counters

60% have a value of 6
40% have a value of 10.

A random sample of 3 counters is drawn from the bag.

a Write down all the possible samples.

b Find the sampling distribution for the median N.

c Find the sampling distribution for the mode M.

Summary of key points

1 A population is a collection of individual items.

2 A sample is a selection of individual members or items from a population.

3 A finite population is one in which each individual member can be given a number.

4 An infinite population is one in which it is impossible to number each member.

5 A sampling unit is an individual member of a population.

6 A sampling frame is a list of sampling units used in practice to represent a population.

7 A statistic is a quantity calculated solely from the observations in a sample.

8 A statistic has a sampling distribution that is defined by giving all possible values of the statistic and the probability of each occurring.

After completing this chapter you should

- understand what is meant by an hypothesis
- understand what is meant by an hypothesis test
- be able to form null and alternative hypotheses for binomial or Poisson distributions
- understand when to use one- and two-tailed tests
- understand what is meant by a critical region
- know what is meant by a significance level
- be able to find critical regions for binomial and Poisson distributions.

Hypothesis testing

Accidents occur on a stretch of road at the rate of 5 a month. The council reduces the speed limit on the road from 40 miles per hour to 30 miles per hour. During the next month the number of accidents is only 3. The council says that the speed reduction is effective. Are the council correct in their claim? After doing this chapter you will be able to find out.

7.1 The concept of an hypothesis test.

- **An hypothesis is a statement made about the value of a population parameter that we wish to test by collecting evidence in the form of a sample.**

- **In a statistical hypothesis test the evidence comes from a sample which is summarized in the form of a statistic called the test statistic.**

We form two hypotheses.

- **The null hypothesis, denoted by H_0, is the hypothesis that we assume to be correct unless proved otherwise.**

- **The alternative hypothesis, denoted by H_1, tells us about the value of the population parameter if our assumption is shown to be wrong.**

Example 1

John wants to see whether a coin is unbiased or whether it is biased towards coming down heads. He tosses the coin 8 times and counts the number of times, X, that it lands head uppermost.

a Describe the test statistic.

b Write down a suitable null hypothesis.

c Write down a suitable alternative hypothesis.

a The test statistic is X (the number of heads) in 8 tosses.

b If the coin is unbiased the probability of a coin landing heads is 0.5 so H_0: $p = 0.5$ is the null hypothesis.

c If the coin is biased towards coming down heads then the probability of landing heads will be greater than 0.5.
H_1: $p > 0.5$ is the alternative hypothesis.

There are other alternative hypotheses.

We might think that the coin is biased to come down tails more often than heads so that the probability of landing heads will be less than 0.5.

In this case H_1: $p < 0.5$ is the alternative hypothesis.

We might think the coin is biased but not know if it is biased towards heads, or whether it is biased towards tails. The probability of landing heads could be smaller or larger than 0.5.

In this case H_1: $p \neq 0.5$ is the alternative hypothesis.

7.2 The significance level of a hypothesis test.

You will have realised that, in Example 1, X is a random variable, and, if the coin is unbiased, theoretically X can take any of the values 0, 1, 2, 3, 4, 5, 6, 7, or 8.
We can write the distribution of X as

You met random variables in book S1. This is the binomial distribution B(8, 0.5).

x	0	1	2	3	4	5	6	7	8
$P(X = x)$	0.004	0.031	0.109	0.219	0.273	0.219	0.109	0.031	0.004

These probabilities can be obtained by using the binomial tables. See Chapter 1.

How many heads do we have to get before we decide we have enough evidence to decide that the coin is biased towards heads?

If we assume that the null hypothesis is true then no value of x between 0 and 8 is impossible.
For example,
there is a $(10.9 + 3.1 + 0.4)\% = 14.4\%$ chance of getting 6, 7 or 8 heads,
a $(3.1 + 0.4)\% = 3.5\%$ chance of getting 7 or 8 heads,
and a 0.4% chance of getting 8 heads.

The decision whether or not to reject the null hypothesis for your observed value has to be based upon the idea that some values of X are unlikely under the null hypothesis and would be better explained by the alternative hypothesis.

We can divide the x values into two regions.

1. The one that contains the values that **collectively** have a small chance of happening under the null hypothesis. We call this region the **critical region**.
2. The rest.

- **The critical region is the range of values of the test statistic that would lead to you rejecting H_0.**

- **The value(s) on the boundary of the critical region are called critical value(s).**

How small is 'small'? To some extent this is a subjective matter, but statisticians generally regard a probability of 5% as being unlikely and a probability of 1% as being very unlikely.

Clearly there is a threshold probability, be it 5% or 1%, and that threshold may vary depending on the nature of the problem.

We call this threshold the **level of significance**.

We use the Greek letter α to represent the level of significance.

Example 2

In Example 1, John wished to see if a coin was unbiased or biased towards coming down heads. He decides that the level of significance of his test will be 5%.

If the random variable X represents the number of heads, what values of X would cause the null hypothesis to be rejected?

$H_0 : p = 0.5$ (the coin is not biased)

$H_1 : p > 0.5$ (the coin is biased towards heads)

The probability of getting 7 or 8 heads is

3.1% + 0.4% = 3.5%

This is less than 5% and either $X = 7$ or $X = 8$ would cause the null hypothesis to be rejected.

The probability of $x = 6$ or greater is 14.4% which is greater than the significance level so $x = 6$ would not cause the null hypothesis to be rejected.

7.3 One- and two-tailed tests.

If the hypothesis test is about a population parameter θ, then we test a null hypothesis H_0 which specifies a particular value for θ, against an alternative hypothesis H_1. It is this alternative hypothesis which will indicate whether the test is one-tailed or two-tailed.

In this book θ will either be the proportion p of a binomial distribution or the mean λ (or μ) of a Poisson distribution. You will meet other parameters in book S3.

- **A one-tailed test looks either for an increase in the value of a parameter or for a decrease in the value of a parameter.**
- **If the null hypothesis is of the form $H_0 : \theta = m$ (for some number m), then a one-tailed test is used when the alternative hypothesis is of the form $H_1 : \theta > m$, (a definite increase in θ), or when it is of the form $H_1 : \theta < m$, (a definite decrease in θ).**
- **A one-tailed test will have a single part to the critical region and one critical value.**

We can show this with two diagrams for the coin tossing experiment. In this case the parameter being tested is the proportion of heads so $\theta = p$ in this case.

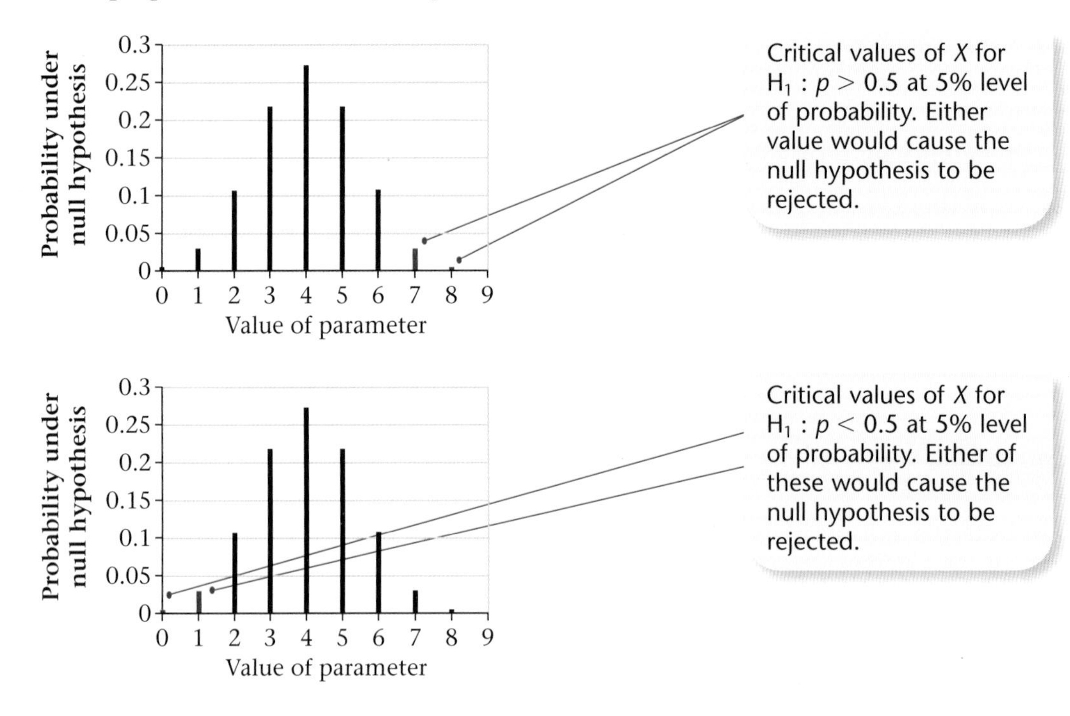

- **A two-tailed test arises when the alternative hypothesis is of the form $H_1 : \theta \neq m$. (That is to say that the alternative could be an increase or a decrease in θ).**

- **If the test is two-tailed, there will be two parts to the critical region and two critical values.**

- **If we want a 5% significance level the usual convention is to allow $2\frac{1}{2}$% at either tail.**

We can again show how this works with a diagram.

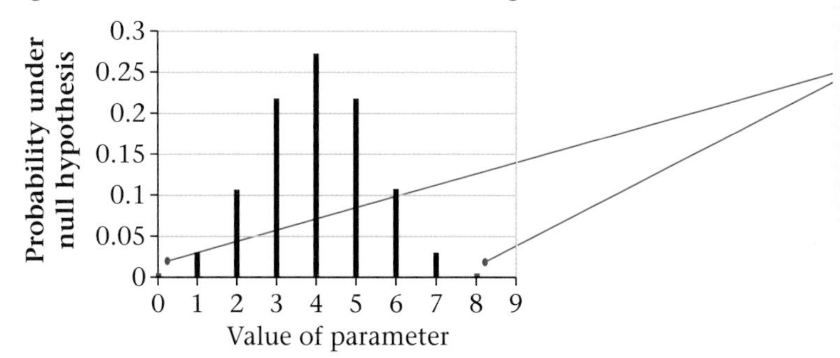

If the observed value of our test statistic x lies in the critical region we have sufficient evidence to reject the null hypothesis H_0.

If the observed value of our test statistic x does not lie in the critical region then we do not have sufficient evidence to reject the null hypothesis H_0.

This does not mean that the null hypothesis is true, rather that it remains a possibility and can be accepted at the moment, but later samples may cause it to be rejected.

We can summarise these test procedures as follows.

1 Identify the population parameter θ (proportion p for a binomial or mean λ (or μ) for a Poisson) that you are going to test.

2 Write down the null (H_0) and alternative (H_1) hypotheses. The alternative hypothesis will determine whether you want a one- or two-tailed test.

3 Specify the significance level α.

4 Find out whether the observed value x of your test statistic falls in the critical region.

Note: you should justify why you have chosen the critical region by giving the appropriate binomial probabilities from tables.

For some given number m

One-tailed tests

$H_0 : \theta = m \quad H_1 : \theta > m \qquad$ Reject if $P(X \geqslant x) \leqslant \alpha$
$H_0 : \theta = m \quad H_1 : \theta < m \qquad$ Reject if $P(X \leqslant x) \leqslant \alpha$

Two-tailed test

$H_0 : \theta = m \quad H_1 : \theta \neq m \qquad$ Reject if $P(X \geqslant x) \leqslant \frac{1}{2}\alpha$ or $P(X \leqslant x) \leqslant \frac{1}{2}\alpha$

5 State your conclusion.
The following points should be addressed.

a Is the result **significant or not**?

b What are the implications in terms of the **context of the original problem**?

Example 3

Accidents used to occur at a road junction at a rate of 6 per month. After a speed limit is placed on the road the number of accidents in the following month is 2. The planners wish to test, at the 5% level of significance, whether or not there has been a decrease in the rate of accidents.

a Suggest a suitable test statistic.

b Write down two suitable hypotheses.

c Explain the condition under which the null hypothesis is rejected.

a The test statistic is the number of accidents in a month.

b $H_0 : \lambda = 6 \qquad H_1 : \lambda < 6$

c If the probability of getting 2 (or fewer) accidents is $\leq 5\%$ the null hypothesis is rejected.

It is important that you identify which is H_0 and which is H_1.

Exercise 7A

1 **a** Describe what is meant by the expression 'a statistical hypothesis'.

b Describe the difference between the null hypothesis and the alternative hypothesis.

c What symbols do we use to denote the null and alternative hypotheses?

2 Dmitri wants to see whether a die is biased towards the value 6.
He throws the die 60 times and counts the number of sixes he gets.

a Describe the test statistic.

b Write down a suitable null hypothesis to test this die.

c Write down a suitable alternative hypothesis to test this die.

3 Shell wants to test to see whether a coin is biased. She tosses the coin 100 times and counts the number of times she gets a head.

a Describe the test statistic.

b Write down a suitable null hypothesis to test this coin.

c Write down a suitable alternative hypothesis to test this coin.

4 Over a long period of time it is found that the mean number of accidents, λ, occurring at a particular crossroads is 4 per month. New traffic lights are installed. Jess decides to test to see whether the rate of occurrence has increased, decreased or changed in any way.

a Describe the test statistic.

b Write down a suitable null hypothesis to test Jess' theory.

c Write down three possible alternative hypotheses to test Jess' theory.

5 In a survey it was found that 4 out of 10 people supported a certain particular political party. Chang wishes to test whether or not there has been a change in the proportion (p) of people supporting the party.

a Write down whether it would be best to use a one-tail test or a two-tail test.

Give a reason for your answer.

b Suggest suitable hypotheses.

6 In a manufacturing process the proportion (p) of faulty articles has been found, from long experience, to be 0.1.
The proportion of faulty articles in the first batch produced by a new process is measured.
The proportion of faulty articles in this batch is 0.09.
The manufacturers wish to test at the 5% level of significance whether or not there has been a reduction in the proportion of faulty articles.

a Suggest a suitable test statistic.

b Write down two suitable hypotheses.

c Explain the condition under which the null hypothesis is rejected.

7 A spinner has 4 sides numbered 1, 2, 3 and 4. Hajdra thinks it is biased to give a one when spun. She spins 5 times and counts the number of times, M, that she gets a 1.

a Describe the test statistic M.

She decides to do a test with a level of significance of 5%.

b What values of M would cause the null hypothesis to be rejected.

7.4 Hypothesis tests for the proportion p of a binomial distribution and hypothesis tests for the mean λ of a Poisson distribution.

Earlier, in Section 7.2, we described the critical region as being the region containing the values that **collectively** have a small chance of happening – typically 5% or less (or 1% or less).

If the observed value x falls in the critical region, then, for one-tailed tests, the $P(X \leqslant x)$ is $\leqslant \alpha$ or the $P(X \geqslant x)$ is $\leqslant \alpha$, depending on the alternative hypothesis.

You should note that it is possible for $P(X = x)$ to be less than say 5% but for x to not be in the critical region.

For example the diagram below shows part of the tail of a B(30, 0.3) distribution.

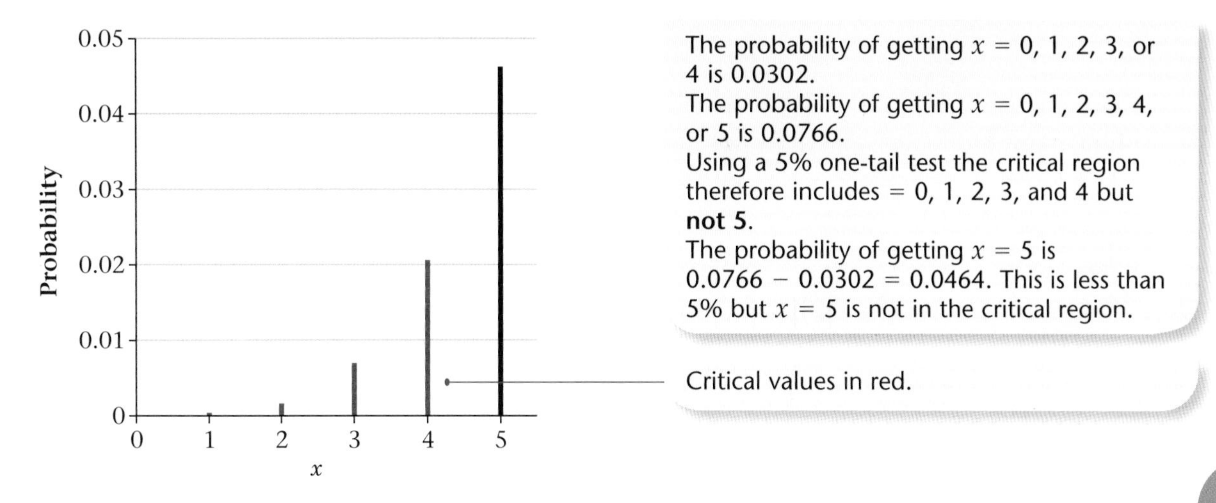

The phrase '**as bad or worse**' is sometimes helpful. We calculate the probability of getting evidence as bad or worse than that we have been presented with to make our judgement.

Testing the observed proportion *p* of a binomial distribution

If the proportion of successful outcomes in the population is p then the test statistic $X \sim B(n, p)$, where n is the number in the sample.

If x is a particular value of X then $P(X \leqslant x)$ may be read directly from the table of the cumulative binomial distribution function, while $P(X \geqslant x)$ can be calculated from $P(X \geqslant x) = 1 - P(X \leqslant x - 1)$.

Example 4

The standard treatment for a particular disease has a $\frac{2}{5}$ probability of success. A certain doctor has undertaken research in this area and has produced a new drug which has been successful with 11 out of 20 patients. The doctor claims that the new drug represents an improvement on the standard treatment.

a Write down suitable null and alternative hypotheses.

b Write down the distribution of the random variable X that represents the number of patients that would be cured if the null hypothesis were true.

c Work out the probability of X taking a value equal to, or greater than, 11.

d If the significance level is to be 5%, does 11 patients out of 20 provide evidence to reject H_0?

a $H_0: p = \frac{2}{5}$ $\quad H_1: p > \frac{2}{5}$

The doctor claims the drug represents an improvement.

b $X \sim B(20, \frac{2}{5})$

c $P(X \geqslant 11) = 1 - P(X \leqslant 10)$

$= 1 - 0.8725$

$= 0.1275$

$= 12.75\%$

d $12.75\% > 5\%$ so there is not enough evidence to reject H_0.

The new drug is no better than the old one.

Sometimes a question is set that does not have values of p that are in the table.

Example 5

Over a long period of time it has been found that in Enrico's restaurant the ratio of non-vegetarian to vegetarian meals is 2 to 1. In Manuel's restaurant in a random sample of 10 people ordering meals, 1 ordered a vegetarian meal. Using a 5% level of significance, test whether or not the proportion of people eating vegetarian meals in Manuel's restaurant is different to that in Enrico's restaurant.

The proportion of people eating vegetarian meals at Enrico's is $\frac{1}{3}$.

Let p be the proportion of people at Manuel's that order a vegetarian meal.

Identify population parameter.

Let X be the number of people in the sample who are eating vegetarian meals.

Test statistic is X.

$H_0 : p = \frac{1}{3}$ $H_1 : p \neq \frac{1}{3}$

Hypotheses. The test will be two-tailed as we are testing if they are different.

Significance level 5%

If H_0 is true $X \sim B(10, \frac{1}{3})$

$$P(X \leq 1) = P(X = 0) + P(X = 1)$$
$$= (\tfrac{2}{3})^{10} + 10(\tfrac{2}{3})^9(\tfrac{1}{3})$$
$$= 0.01734\ldots + 0.08670\ldots$$
$$= 0.10404\ldots$$
$$= 0.104 \text{ (3 s.f.)}$$

This time we have to calculate the probabilities.

$0.104 > 0.025$

We use 0.025 because the test is two-tailed.

There is insufficient evidence to reject H_0.

There is no evidence that proportion of vegetarian meals at Manuel's restaurant is different to Enrico's.

Conclusion and what it means in context.

Testing the observed rate λ of a Poisson distribution

When testing the mean or rate of a Poisson distribution the table for the Poisson cumulative distribution function may be used.

Example 6

Accidents used to occur at a certain road junction at the rate of 6 per month. The residents petitioned for traffic lights. In the month after the lights were installed there was only 1 accident. Does this give sufficient evidence that the lights have reduced the number of accidents? Use a 5% level of significance.

Let the random variable X represent the number of accidents in a month and λ represent the rate per month.

Test statistic X.
Identify population parameter.

$H_0 : \lambda = 6$ $H_1 : \lambda < 6$

Hypotheses. The test will be one-tailed.

Under H_0 $X \sim Po(6)$

Significance level 5%

$P(X \leq 1 \mid \lambda = 6) = 0.0174$ — Find the probabilities.

$0.0174 < 0.05$ (5%) — We use 0.05 because the test is one-tailed.

There is sufficient evidence to reject H_0.

Lights may have reduced the number of accidents.

Again questions may be asked which cannot be answered using the tables.

Example 7

Over a long period of time, Jessie found that the bus taking her to school was late at the rate of 6.7 times per month. In the month following the start of the new summer bus schedules, Jessie finds that her bus is late twice. Assuming that the number of times the bus is late has a Poisson distribution, test at the 1% level of significance, whether or not the new schedules have in fact decreased the number of times the bus is late.

Let λ be the average number of times late.

$X \sim Po(\lambda)$

$H_0 : \lambda = 6.7 \quad H_1 : \lambda < 6.7$

Level of significance 1%

$$P(X \leq 2 \mid \lambda = 6.7)$$
$$= P(X = 0) + P(X = 1) + P(X = 2)$$
$$= e^{-6.7}\left(1 + 6.7 + \frac{6.7^2}{2}\right)$$
$$= e^{-6.7}(1 + 6.7 + 22.445)$$
$$= e^{-6.7}(30.145)$$
$$= 0.0371 \text{ (3 s. f.)}$$

$0.0371 > 0.01$

There is insufficient evidence to reject H_0.

The new schedule has not decreased the number of times the bus is late.

Exercise 7B

For each of the questions 1 to 7 carry out the following tests using the binomial distribution where the random variable, *X*, represents the number of successes.

1 $H_0 : p = 0.25$; $H_1 : p > 0.25$; $n = 10$, $x = 5$ and using a 5% level of significance.

2 $H_0 : p = 0.40$; $H_1 : p < 0.40$; $n = 10$, $x = 1$ and using a 5% level of significance.

3 $H_0 : p = 0.30$; $H_1 : p > 0.30$; $n = 20$, $x = 10$ and using a 5% level of significance.

4 $H_0 : p = 0.45$; $H_1 : p < 0.45$; $n = 20$, $x = 3$ and using a 1% level of significance.

5 $H_0 : p = 0.50$; $H_1 : p \neq 0.50$; $n = 30$, $x = 10$ and using a 5% level of significance.

6 $H_0 : p = 0.28$; $H_1 : p < 0.28$; $n = 20$, $x = 2$ and using a 5% level of significance.

7 $H_0 : p = 0.32$; $H_1 : p > 0.32$; $n = 8$, $x = 7$ and using a 5% level of significance.

For each of the questions 8 to 10 carry out the following tests using the Poisson distribution where λ represents its mean.

8 $H_0 : \lambda = 8$; $H_1 : \lambda < 8$; $x = 3$ and using a 5% level of significance.

9 $H_0 : \lambda = 6.5$; $H_1 : \lambda < 6.5$; $x = 2$ and using a 1% level of significance.

10 $H_0 : \lambda = 5.5$; $H_1 : \lambda > 5.5$; $x = 8$ and using a 5% level of significance.

11 The manufacturer of 'Supergold' margarine claims that people prefer this to butter. As part of an advertising campaign he asked 5 people to taste a sample of 'Supergold' and a sample of butter and say which they prefer. Four people chose 'Supergold'. Assess the manufacturer's claim in the light of this evidence. Use a 5% level of significance.

12 I tossed a coin 20 times and obtained a head on 6 occasions. Is there evidence that the coin is biased? Use a 5% two-tailed test.

13 A die used in playing a board game is suspected of not giving the number 6 often enough. During a particular game it was rolled 12 times and only one 6 appeared. Does this represent significant evidence, at the 5% level of significance, that the probability of a 6 on this die is less than $\frac{1}{6}$?

14 The success rate of the standard treatment for patients suffering from a particular skin disease is claimed to be 68%.

a In a sample of n patients, X is the number for which the treatment is successful. Write down a suitable distribution to model X. Give reasons for your choice of model.

A random sample of 10 patients receives the standard treatment and in only 3 cases was the treatment successful. It is thought that the standard treatment was not as effective as it is claimed.

b Test the claim at the 5% level of significance.

15 Every year a statistics teacher takes her class out to observe the traffic passing the school gates during a Tuesday lunch hour. Over the years she has established that the average number of lorries passing the gates in a lunch hour is 7.5. During the last 12 months a new bypass has been built and the number of lorries passing the school gates in this year's experiment was 4. Test, at the 5% level of significance, whether or not the mean number of lorries passing the gates during a Tuesday lunch hour has been reduced.

16 Over a long period, John has found that the bus taking him to school arrives late on average 9 times per month. In the month following the start of the new summer schedule the bus arrives late 13 times. Assuming that the number of times the bus is late has a Poisson distribution, test, at the 5% level of significance, whether the new schedules have in fact increased the number of times on which the bus is late. State clearly your null and alternative hypotheses.

7.5 Hypothesis tests for the proportion p of a binomial distribution and hypothesis tests for the rate λ of a Poisson distribution using critical regions.

Sometimes it is helpful to consider what is the least (or greatest) value of the test statistic that would provide sufficient evidence for rejecting H_0. These are called the critical values.

Nearly all other hypothesis tests are done by finding the critical values and thus the critical regions. If you have a choice it is easier not to use critical values, but examination questions often ask for the critical regions to be found and used.

Testing the proportion p of a binomial distribution

Example 8

A psychologist is attempting to help a student improve his short term memory. One of the tests the psychologist uses is to present the student with a tray of 10 objects and let him look at them for one minute before taking the tray away and asking the student to write down as many of the objects as he can. Over a period of several weeks the psychologist ascertains that the proportion, p, of objects that the student remembers is 0.35. The student has just been on a long adventure holiday and the psychologist is interested to see if there has been any change in p. Find the critical values for a two-tailed test using a 5% level of significance.

Let p represent the proportion of objects the student remembers.

Significance level 5%

$H_0: p = 0.35 \quad H_1: p \neq 0.35$

Assuming H_0 is true then $X \sim B(10, 0.35)$

This is a two-tailed test so let c_1 and c_2 be the two critical values.

Two-tailed as psychologist is looking for a change.

Then $P(X \leqslant c_1) \leqslant 0.025$ and $P(X \geqslant c_2) \leqslant 0.025$

From the binomial table $P(X \leqslant 0) = 0.0135$

and $P(X \leqslant 1) = 0.0860$

We look for two adjacent values of X with probabilities either side of 0.025.

so the value of c_1 is 0 since $P(X \leqslant 0) = 0.0135 < 0.025$.

From the binomial table $P(X \geqslant 7) = 1 - P(X \leqslant 6) = 1 - 0.9740 = 0.0260$

and $P(X \geqslant 8) = 1 - P(X \leqslant 7) = 1 - 0.9952 = 0.0048$

so the value of $c_2 = 8$ since $P(X \geqslant 8) = 0.0048 < 0.025$.

These lines should be included to justify your choice and show which values have been compared.

The critical region for X is $X = 0$ or $X \geqslant 8$.

Because we are dealing with a discrete distribution the values of X have to be taken to the nearest integer. The test that results will usually have a probability of rejecting H_0 which is lower than the intended significance level of the test.

■ **The actual level of significance of the test is the probability of rejecting H_0.**

The actual significance level is $P(X \leqslant c)$ or $P(X \geqslant c)$ for a one-tailed test or $P(X \leqslant c_1) + P(X \geqslant c_2)$ for a two-tailed test.

In Example 8

$$\begin{aligned} P(\text{rejecting } H_0) &= P(X = 0 \mid p = 0.35) + P(X \geqslant 8) \\ &= 0.0135 + 0.0048 \\ &= 0.0183 \end{aligned}$$

and this is not very close to the 5% significance level required.

It is the usual practice, if a 5% two-tailed test is required, to use $2\frac{1}{2}\%$ at each end, but sometimes we make the tails as close to $2\frac{1}{2}\%$ as possible, rather than $\leqslant 2\frac{1}{2}\%$.

Example 9

A psychologist is attempting to help a student improve his short term memory. One of the tests the psychologist uses is to present the student with a tray of 10 objects and let him look at them for one minute before taking the tray away and asking the student to write down as many of the objects as he can. Over a period of several weeks the psychologist ascertains that the proportion, p, of objects that the student remembers is 0.35. The student has just been on a long adventure holiday and the psychologist is interested to see if there has been any change in p. Find the critical values for a two-tailed test using as close as possible to $2\frac{1}{2}\%$ level of significance at each tail.

Let p represent the proportion of objects the student remembers.

Significance level 5%

$H_0: p = 0.35 \quad H_1: p \neq 0.35$

Assuming H_0 is true then $X \sim B(10, 0.35)$

This is a two-tailed test so let c_1 and c_2 be the two critical values.

This approach should only be used when specifically told to use it.

From the binomial table $P(X \leqslant 0) = 0.0135$

and $P(X \leqslant 1) = 0.0860$

so the value of c_1 is 0.

From the binomial table $P(X \geqslant 6) = 1 - P(X \leqslant 5) = 1 - 0.9051 = 0.0849$

$P(X \geqslant 7) = 1 - P(X \leqslant 6) = 1 - 0.9740 = 0.0260$

and $P(X \geqslant 8) = 1 - P(X \leqslant 7) = 1 - 0.9952 = 0.0048$

so the value of $c_2 = 7$

The critical region for X is $X = 0$ or $X \geqslant 7$.

The actual level of significance is now $0.0260 + 0.0135 = 0.0395$ or 3.95%.

Testing the mean λ of a Poisson distribution

When testing the mean of a Poisson distribution the table for the Poisson cumulative distribution function may be used.

Example 10

An office finds that over a long time incoming telephone calls from customers occur at a rate of 0.325 per minute.

They believe that the number of calls has increased recently. To test this, the number of incoming calls during a random 20-minute interval is recorded.

a Find the critical region for a two-tailed test of the hypothesis that the number of incoming calls occur at the rate of 0.325 per minute. The probability in both tails should be as close to 2.5% as possible.

b Write down the actual significance level of the above test.

Later the office runs an advertising campaign and records 1 call in a 10-minute interval.

c Test, at the 5% level of significance, whether or not there is evidence that the rate of incoming calls has decreased.

a If the rate of calls is 0.325 calls per minute then in 20 minutes you would expect

$20 \times 0.325 = 6.5$ calls

Let X be the number of telephone calls then $X \sim \text{Po}(6.5)$

$H_0 : \lambda = 6.5 \quad H_1 : \lambda \neq 6.5$

Significance level 5% (as near as possible 2.5% at each tail)

This is a two-tailed test so let c_1 and c_2 be the critical values.

From the tables $\quad P(X \leqslant 2 \mid \lambda = 6.5) = 0.0430$

and $\quad P(X \leqslant 1 \mid \lambda = 6.5) = 0.0113$

so $c_1 = 1$ (0.0113 is nearer to 0.025 than is 0.0430)

From the tables $P(X \geqslant 12 \mid \lambda = 6.5) = 1 - P(X \leqslant 11 \mid \lambda = 6.5)$

$= 1 - 0.9661 = 0.0339$

and $\quad P(X \geqslant 13 \mid \lambda = 6.5) = 1 - P(X \leqslant 12 \mid \lambda = 6.5)$

$= 1 - 0.9840 = 0.0160$

so $c_2 = 12$ (0.0339 is nearer to 0.025 than is 0.0160)

The critical region will be $X \leqslant 1$ and $X \geqslant 12$

b The actual significance level will be $0.0113 + 0.0339 = 0.0452$

c If there are 0.325 calls per minute there will be $10 \times 0.325 = 3.25$ calls in 10 minutes.

This will be a Po(3.25) distribution.

$H_0: \lambda = 3.25 \; H_1: \lambda < 3.25$

This is a one-tail test. Significance level 5%

The hypotheses can be stated in terms of the **rate**.

In terms of rate the hypotheses would be
$H_0 : \lambda = 0.325$
$H_1 : \lambda \neq 0.325$

If you had not used nearest to $2\frac{1}{2}$% the actual significance level would have been 0.0273.

Let c be the critical value.

$P(X = 0 \mid \lambda = 3.25) = e^{-3.25} = 0.0388$

$P(X = 1 \mid \lambda = 3.25) = 0.0388 \times 3.25 = 0.1260$

so $c = 0$ and the critical region is $X = 0$ since $P(X = 0) = 0.0388 < 0.05$

$X = 1$ is not in the critical region so there is insufficient evidence to reject H_0.

There is no evidence that the rate of telephone calls has decreased.

Using approximations

Whether you are using critical regions or not, with both the binomial and Poisson distributions, the numbers in the sample could be large in practice. This makes the calculations very difficult. In these cases a normal approximation could be considered if the conditions are suitable. If you have a binomial with n large and p small then you can use a Poisson as an approximation for a binomial distribution. The following examples illustrate these.

Example 11

A shop sells grass mowers at the rate of 10 per week. In an attempt to increase sales, the price was reduced for a six-week period. During this period a total of 75 mowers were sold.

Using a 5% level of significance, test whether or not there is evidence that the average number of sales per week has increased during this six-week period.

$H_0 : \lambda = 10,\ H_1 : \lambda > 10$

Let Y represent the number sold in a six-week period then,

under H_0, $Y \sim Po(60)$

$P(Y \geqslant 75) \approx P(W > 74.5)$ where $W \sim N(60, 60)$

Don't forget the 0.5 continuity correction.

$\approx P\left(Z > \dfrac{74.5 - 60}{\sqrt{60}}\right) = P(Z > 1.87) = 0.0307$

$0.0307 < 0.05$, therefore reject H_0.

There is evidence that the sales per week have increased.

Example 12

A manager thinks that his sales staff make a sale to 45% of customers entering their shop. He randomly selects 100 customers. Of the 100 customers, 35 were sold something.

Using a suitable approximation test, at the 5% level of significance, whether or not what the manager thinks is justified.

$H_0 : p = 0.45$, $H_1 : p \neq 0.45$

Here n is large and p is near 0.5 so a normal approximation can be used.

$B(100, 0.45) \approx \sim N(45, 24.75)$

$P(X \leqslant 35) \approx P(W < 35.5)$

$$P(W < 35.5) = P\left(Z < \frac{35.5 - 45}{\sqrt{24.75}}\right)$$
$$= P(Z < -1.91)$$
$$= 1 - 0.9719$$
$$= 0.0281$$

$0.0281 > 0.025$, therefore accept H_0.

There is no evidence that the manager is wrong.

$\approx$ means approximately equal.
$\sim$ means the distribution.
So this line reads as:
B(100, 0.45) is approximately equal to the distribution N(45, 24.75).

Example 13

During an influenza epidemic, 4% of the population of a large city was affected on a given day. The manager of a factory that employs 100 people found that 12 of his employees were absent, claiming to have influenza.

a Using a 5% significance level, find the critical region that would enable the manager to test whether or not there is evidence that the percentage of people having influenza at his factory was greater than that of the large city.

b State the conclusion the manager came to giving a reason for your answer.

a $H_0 : p = 0.04 \quad H_1 : p > 0.04$

Let X be the number of absentees.

Under H_0 $\quad X \sim B(100, 0.04) \approx \sim Po(4)$

$P(X \geqslant 9) = 1 - P(X \leqslant 8) = 1 - 0.9786 = 0.0214$

$P(X \geqslant 8) = 1 - P(X \leqslant 7) = 1 - 0.9489 = 0.0511$

Critical region $X \geqslant 9$ since $P(X \geqslant 9) = 0.0214 < 0.05$

b Since $12 > 9$ the manger concluded that the percentage of people having influenza at his factory was larger than that of the city.

Exercise 7C

For each of the questions 1 to 6 find the critical region for the test statistic X representing the number of successes. Assume a binomial distribution.

1 $H_0 : p = 0.20$; $H_1 : p > 0.20$; $n = 10$, using a 5% level of significance.

2 $H_0 : p = 0.15$; $H_1 : p < 0.15$; $n = 20$, using a 5% level of significance.

3 $H_0 : p = 0.40$; $H_1 : p \neq 0.40$; $n = 20$, using a 5% level of significance (2.5% at each tail).

4 $H_0 : p = 0.18$; $H_1 : p < 0.18$; $n = 20$, using a 1% level of significance.

5 $H_0 : p = 0.63$; $H_1 : p > 0.63$; $n = 10$, using a 5% level of significance.

6 $H_0 : p = 0.22$; $H_1 : p \neq 0.22$; $n = 10$, using a 1% level of significance (0.005 at each tail).

For each of the questions 7 to 9 find the critical region for the test statistic X given that X has a Po(λ) distribution.

7 $H_0 : \lambda = 4$; $H_1 : \lambda > 4$; using a 5% level of significance.

8 $H_0 : \lambda = 9$; $H_1 : \lambda < 9$; using a 1% level of significance.

9 $H_0 : \lambda = 3.5$; $H_1 : \lambda < 3.5$; using a 5% level of significance.

10 A seed merchant usually kept her stock in carefully monitored conditions. After the Christmas holidays one year she discovered that the monitoring system had broken down and there was a danger that the seed might have been damaged by frost. She decided to check a sample of 10 seeds to see if the proportion p that germinates had been reduced from the usual value of 0.85. Find the critical region for a one-tailed test using a 5% level of significance.

11 The national proportion of people experiencing complications after having a particular operation in hospitals is 20%. A particular hospital decides to take a sample of size 20 from their records.

a State all the possible numbers of patients with complications that would cause them to decide that their proportion of complications differs from the national figure at the 5% level of significance, ensuring that the probability in each tail is as near to 2.5% as possible.

The hospital finds that out of 20 such operations, 8 of their patients experienced complications.

b Find critical regions, at the 5% level of significance, to test whether or not their proportion of complications differs from the national proportion. The probability in each tail should be as near 2.5% as possible.

c State the actual significance level of the above test.

12 Over a number of years the mean number of hurricanes experienced in a certain area during the month of August is 4. A scientist suggests that, due to global warming, the number of hurricanes will have increased, and proposes to do a hypothesis test based on the number of hurricanes this year.

a Suggest suitable hypotheses for this test.

b Find to what level the number of hurricanes must increase for the null hypothesis to be rejected at the 5% level of significance.

c The actual number of hurricanes this year was 8. What conclusion did the scientist come to?

13 An estate agent usually sells properties at the rate of 10 per week.
During a recession, when money was less available, over an eight-week period he sold 55 properties.
Using a suitable approximation test, at the 5% level of significance, whether or not there is evidence that the weekly rate of sales decreased.

14 A manager thinks that 20% of his workforce are absent for at least one day each month.
He chooses 100 workers at random and finds that in the last month 2 had been absent for at least one day.
Using a suitable approximation test, at the 5% level of significance, whether or not this provides evidence that the percentage of workers that are absent for at least 1 day per month is less than 20%.

Mixed exercise 7D

1 Mai commutes to work five days a week on a train. She does two journeys a day.
Over a long period of time she finds that the train is late 20% of the time.
A new company takes over the train service Mai uses. Mai thinks that the service will be late more often. In the first week of the new service the train is late 3 times.
You may assume that the number of times the train is late in a week has a binomial distribution.
Test, at the 5% level of significance, whether or not there is evidence that there is an increase in the number of times the train is late. State your hypothesis clearly.

2 Over a long period of time it was observed that the mean number of lorries passing a hospital was 7.5 every 10 minutes.
A new by-pass was built that avoided the hospital. In a survey after the by-pass was opened, it was found that in one particular week the mean number of lorries passing the hospital was 4 every 10 minutes. It is decided that a significance test will be done to test whether or not the mean number of lorries passing the hospital has changed.

a State whether a one- or two-tailed test will be needed. Give a reason for your answer.

b Write down the name of the distribution that will be tested. Give a reason for your choice.

c Carry out the significance test at the 5% level of significance.

3 A marketing company claims that Chestly cheddar cheese tastes better than Cumnauld cheddar cheese.
Five people chosen at random as they entered a supermarket were asked to say which they preferred. Four people preferred Chestly cheddar cheese.
Test, at the 5% level of significance, whether or not the manufacturers claim is true. State your hypothesis clearly.

4 In 2006 and 2007 much of Greebe suffered earth tremors at a rate of 5 per month.
A survey was done in the first two months of 2008 and 13 tremors were recorded.
Stating your hypothesis clearly test, at the 10% level of significance, whether or not there is evidence to suggest the rate of earth tremors has increased.

5 Historical information finds that nationally 30% of cars fail a brake test.

a Give a reason to support the use of a binomial distribution as a suitable model for the number of cars failing a brake test.

b Find the probability that, of 5 cars taking the test, all of them pass the brake test.

A garage decides to conduct a survey of their cars. A randomly selected sample of 10 of their cars is tested. Two of them fail the test.

c Test, at the 5% level of significance, whether or not there is evidence to support the suggestion that cars in this garage fail less than the national average.

6 **a** Explain what you understand by an hypothesis test.

During a garden fete cups of tea are thought to be sold at a rate of 2 every minute. To test this, the number of cups of tea sold during a random 30-minute interval is recorded.

b State one reason why the sale of cups of tea can be modelled by a Poisson distribution.

c Find the critical region for a two-tailed hypothesis that the number of cups of tea sold occurs at a rate of 2 every minute. The probability in each tail should be as close to 2.5% as possible.

d Write down the actual significance level of the above test.

7 The probability that Jacinth manages to hit a coconut on the coconut shy at a fair is 0.4. She decides to practise at home. After practising she thinks that the practising has helped her to improve. After practising Jacinth is going to the fair and will have 20 throws.

a Find the critical region for an hypothesis test at the 5% level of significance.

After practising, Jacinth hits the coconut 11 times.

b Determine whether or not there is evidence that practising has helped Jacinth improve. State your hypothesis clearly.

8 The proportion of defective articles in a certain manufacturing process has been found from long experience to be 0.1.
A random sample of 50 articles was taken in order to monitor the production. The number of defective articles was recorded.

a Using a 5% level of significance, find the critical regions for a two-tailed test of the hypothesis that 1 in 10 articles has a defect. The probability in each tail should be as near 2.5% as possible.

b State the actual significance level of the above test.

Another sample of 20 articles was taken at a later date. Four articles were found to be defective.

c Test at the 10% significance level, whether or not there is evidence that the proportion of defective articles has increased. State your hypothesis clearly.

9 It is claimed that 50% of women use Oriels powder. In a random survey of 20 women 12 said they did not use Oriels powder.
Test at the 5% significance level, whether or not there is evidence that the proportion of women using Oriels powder is 0.5. State your hypothesis carefully.

10 A large caravan company hires caravans out for a week at a time. During winter the mean number of caravans hired is 6 per week.

a Calculate the probability that in one particular week in winter the company will hire out exactly 4 caravans.

The company decides to reduce prices in winter and do extra advertising. This results in the mean number of caravans being hired out rising to 11 per week.

b Test, at the 5% significance level, whether or not the proportion of caravans hired out has increased. State your hypothesis clearly.

11 The manager of a superstore thinks that the probability of a person buying a certain make of computer is only 0.2.
To test whether this hypothesis is true the manager decides to record the make of computer bought by a random sample of 50 people who bought a computer.

a Find the critical region that would enable the manager to test whether or not there is evidence that the probability is different from 0.2. The probability of each tail should be as close to 2.5% as possible.

b Write down the significance level of this test.

15 people buy that certain make.

c Carry out the significance test. State your hypothesis clearly.

12 At one stage of a water treatment process the number of particles of foreign matter per litre present in the water has a Poisson distribution with mean 10. The water then enters a filtration bed which should extract 75% of foreign matter.
The manager of the treatment works orders a study into the effectiveness of this filtration bed. Twenty samples, each of 1 litre, are taken from the water and 64 particles of foreign matter are found.
Using a suitable approximation test, at the 5% level of significance, whether or not there is evidence that the filter bed is failing to work properly. *E*

13 A shop finds that it sells jars of onion marmalade at the rate of 10 per week. During a television cookery program, onion marmalade is used in a recipe.
Over the next six weeks the shop sells 84 jars of onion marmalade. Using a suitable approximation test at the 5% significance level whether or not there is evidence that the rate of sales after the television program has increased as a result of the television program.

14 A manufacturer produces large quantities of patterned plates. It is known from previous records that 6% of the plates will be seconds because of flaws in the patterns.
To verify that the production process is not getting worse the manager takes a sample of 150 plates and finds that 15 have flaws in their patterns. Use a suitable approximation to test, at the 5% significance level, whether or not the process is getting worse.

15 Jack grows apples. Over a period of time he finds that the probability of an apple being below the size required by a supermarket is 0.45.
He has recently set another orchard using a different variety of apple. A sample of 200 of this new type of apple had 60 rejected as being undersize.
Use a suitable approximation to test, at the 5% significance level, whether or not the new variety of apple is better than the old type of apple.

Summary of key points

1 An **hypothesis test** is a mathematical procedure to examine a value of a population parameter proposed by the null hypothesis H_0, compared to the alternative hypothesis H_1.

2 In an hypothesis test the evidence comes from a sample which is summarised in the form of a **test statistic**.

3 The **critical region** is the range of values of a test statistic that would lead you to reject H_0.

4 The boundary value(s) of a critical region is (are) called the **critical value(s)**.

5 A **one-tailed test** looks either for an increase **or** for a decrease in a parameter, and has a single critical value.

6 A **two-tailed test** looks for both an increase **and** a decrease in a parameter, and has two critical values.

7 The actual significance level of a test is the probability of rejecting H_0.

Review Exercise

1 The random variable X is uniformly distributed over the interval $[-1, 5]$.

a Sketch the probability density function, $f(x)$, of X.

Find

b $E(X)$,

c $Var(X)$,

d $P(-0.3 < X < 3.3)$. *E*

2 A bag contains a large number of coins. Half of them are 1p coins, one third are 2p coins and the remainder are 5p coins.

a Find the mean and variance of the value of the coins.

A random sample of 2 coins is chosen from the bag.

b List all the possible samples that can be drawn.

c Find the sampling distribution of the mean value of these samples. *E*

3 A teacher thinks that 20% of the pupils in a school read the Deano comic regularly.

He chooses 20 pupils at random and finds 9 of them read Deano.

a i Test, at the 5% level of significance, whether or not there is evidence that the percentage of pupils that read Deano is different from 20%. State your hypotheses clearly.

ii State all the possible numbers of pupils that read Deano from a sample of size 20 that will make the test in part **a i** significant at the 5% level.

The teacher takes another 4 random samples of size 20 and they contain 1, 3, 1 and 4 pupils that read Deano.

b By combining all 5 samples and using a suitable approximation test, at the 5% level of significance, whether or not this provides evidence that the percentage of pupils in the school that read Deano is not 20%.

c Comment on your results for the tests in part **a** and part **b**. *E*

4 The continuous random variable X is uniformly distributed over the interval $[2, 6]$.

a Write down the probability density function f(x).

Find

b E(X),

c Var(X),

d the cumulative distribution function of X, for all x,

e $P(2.3 < X < 3.4)$. 

5 The random variable X is the number of misprints per page in the first draft of a novel.

a State two conditions under which a Poisson distribution is a suitable model for X.

The number of misprints per page has a Poisson distribution with mean 2.5. Find the probability that

b a randomly chosen page has no misprints,

c the total number of misprints on 2 randomly chosen pages is more than 7.

The first chapter contains 20 pages.

d Using a suitable approximation find, to 2 decimal places, the probability that the chapter will contain fewer than 40 misprints. **E**

6 Explain what you understand by

a a sampling unit,

b a sampling frame,

c a sampling distribution. **E**

7 A drugs company claims that 75% of patients suffering from depression recover when treated with a new drug.

A random sample of 10 patients with depression is taken from a doctor's records.

a Write down a suitable distribution to model the number of patients in this sample who recover when treated with the new drug.

Given that the claim is correct,

b find the probability that the treatment will be successful for exactly 6 patients.

The doctor believes that the claim is incorrect and the percentage who will recover is lower. From her records she took a random sample of 20 patients who had been treated with the new drug. She found that 13 had recovered.

c Stating your hypotheses clearly, test, at the 5% level of significance, the doctor's belief.

d From a sample of size 20, find the greatest number of patients who need to recover for the test in part **c** to be significant at the 1% level. **E**

8 **a** Explain what you understand by a census.

Each cooker produced at GT Engineering is stamped with a unique serial number. GT Engineering produces cookers in batches of 2000. Before selling them, they test a random sample of 5 to see what electric current overload they will take before breaking down.

b Give one reason, other than to save time and cost, why a sample is taken rather than a census.

c Suggest a suitable sampling frame from which to obtain this sample.

d Identify the sampling units. **E**

9 Dhriti grows tomatoes. Over a period of time, she has found that there is a probability 0.3 of a ripe tomato having a diameter greater than 4 cm. She decides to try a new fertiliser. In a random sample of 40 ripe tomatoes, 18 have a diameter greater than 4 cm. Dhriti claims that the new fertiliser has increased the probability of a ripe tomato being greater than 4 cm in diameter.

Test Dhriti's claim at the 5% level of significance. State your hypotheses clearly. *E*

10 The probability that a sunflower plant grows over 1.5 metres high is 0.25. A random sample of 40 sunflower plants is taken and each sunflower plant is measured and its height recorded.

a Find the probability that the number of sunflower plants over 1.5 m high is between 8 and 13 (inclusive) using

i a Poisson approximation,

ii a normal approximation.

b Write down which of the approximations used in part **a** is a more accurate estimate of the probability. You must give a reason for your answer. *E*

11 **a** Explain what you understand by

i an hypothesis test,

ii a critical region.

During term time, incoming calls to a school are thought to occur at a rate of 0.45 per minute. To test this, the number of calls during a random 20-minute interval is recorded.

b Find the critical region for a two-tailed test of the hypothesis that the number of incoming calls occurs at a rate of 0.45 per 1-minute interval. The probability in each tail should be as close to 2.5% as possible.

c Write down the actual significance level of the above test.

In the school holidays, 1 call occurs in a 10-minute interval.

d Test, at the 5% level of significance, whether or not there is evidence that the rate of incoming calls is less during the school holidays than in term time. *E*

12 A string AB of length 5 cm is cut, in a random place C, into two pieces. The random variable X is the length of AC.

a Write down the name of the probability distribution of X and sketch the graph of its probability density function.

b Find the values of $\mathrm{E}(X)$ and $\mathrm{Var}(X)$.

c Find $\mathrm{P}(X > 3)$.

d Write down the probability that AC is 3 cm long *E*

13 Bacteria are randomly distributed in a river at a rate of 5 per litre of water. A new factory opens and a scientist claims it is polluting the river with bacteria. He takes a sample of 0.5 litres of water from the river near the factory and it contains 7 bacteria. Stating your hypotheses clearly, test his claim at the 5% level of significance. *E*

14 A bag contains a large number of coins

75% are 10p coins,

25% are 5p coins.

A random sample of 3 coins is drawn from the bag.

Find the sampling distribution for the median of the values of the 3 selected coins. *E*

15 Linda regularly takes a taxi to work five times a week. Over a long period of time she finds the taxi is late once a week. The taxi firm changes her driver and Linda thinks the taxi is late more often. In the first week with the new driver, the taxi is late 3 times.

You may assume that the number of times a taxi is late in a week has a binomial distribution.

Test, at the 5% level of significance, whether or not there is evidence of an increase in the proportion of times the taxi is late. State your hypotheses clearly.

16 **a** **i** Write down two conditions for $X \sim B(n, p)$ to be approximated by a normal distribution $Y \sim N(\mu, \sigma)$.

ii Write down the mean and variance of this normal approximation in terms of n and p.

A factory manufactures 2000 DVDs every day. It is known that 3% of DVDs are faulty.

b Using a normal approximation, estimate the probability that at least 40 faulty DVDs are produced in one day.

The quality control system in the factory identifies and destroys every faulty DVD at the end of the manufacturing process. It costs £0.70 to manufacture a DVD and the factory sells non-faulty DVDs for £11.

c Find the expected profit made by the factory per day.

17 **a** Define a statistic.

A random sample $X_1, X_2, \ldots, X_n$ is taken from a population with unknown mean μ.

b For each of the following state whether or not it is a statistic.

i $\dfrac{X_1 + X_4}{2}$,

ii $\dfrac{\Sigma X^2}{n} - \mu^2$.

18 For a particular type of plant 45% have white flowers and the remainder have coloured flowers. Gardenmania sells plants in batches of 12. A batch is selected at random.

Calculate the probability this batch contains

a exactly 5 plants with white flowers,

b more plants with white flowers than coloured ones.

Gardenmania takes a random sample of 10 batches of plants.

c Find the probability that exactly 3 of these batches contain more plants with white flowers than coloured ones.

Due to an increasing demand for these plants by large companies, Gardenmania decides to sell them in batches of 50.

d Use a suitable approximation to calculate the probability that a batch of 50 plants contains more than 25 plants with white flowers.

19 **a** State the condition under which the normal distribution may be used as an approximation to the Poisson distribution.

b Explain why a continuity correction must be incorporated when using the normal distribution as an approximation to the Poisson distribution.

A company has yachts that can only be hired for a week at a time. All hiring starts on a Saturday. During the winter the mean number of yachts hired per week is 5.

c Calculate the probability that fewer than 3 yachts are hired on a particular Saturday in winter.

During the summer the mean number of yachts hired per week increases to 25. The company has only 30 yachts for hire.

d Using a suitable approximation find the probability that the demand for yachts cannot be met on a particular Saturday in summer.

In the summer there are 16 Saturdays on which a yacht can be hired.

e Estimate the number of Saturdays in the summer that the company will not be able to meet the demand for yachts. **E**

20 The continuous random variable X is uniformly distributed over the interval $\alpha < x < \beta$.

a Write down the probability density function of X, for all x.

b Given that $E(X) = 2$ and $P(X < 3) = \frac{5}{8}$, find the value of α and the value of β.

A gardener has wire cutters and a piece of wire 150 cm long which has a ring attached at one end. The gardener cuts the wire, at a randomly chosen point, into 2 pieces. The length, in cm, of the piece of wire with the ring on it is represented by the random variable X. Find

c $E(X)$,

d the standard deviation of X,

e the probability that the shorter piece of wire is at most 30 cm long. **E**

21 Past records from a large supermarket show that 20% of people who buy chocolate bars buy the family size bar. On one particular day a random sample of 30 people was taken from those that had bought chocolate bars and 2 of them were found to have bought a family size bar.

a Test, at the 5% significance level, whether or not the proportion p of people who bought a family size bar of chocolate that day had decreased. State your hypotheses clearly.

The manager of the supermarket thinks that the probability of a person buying a gigantic chocolate bar is only 0.02. To test whether this hypothesis is true the manager decides to take a random sample of 200 people who bought chocolate bars.

b Find the critical region that would enable the manager to test whether or not there is evidence that the probability is different from 0.02. The probability of each tail should be as close to 2.5% as possible.

c Write down the significance level of this test. **E**

Examination practice paper

1 **a** Explain briefly what you understand by

i a sampling frame,

ii a statistic. (3)

A random sample $X_1, X_2, \ldots, X_n$ is taken from a population with unknown mean μ.

b For each of the following state whether or not it is a statistic. Give a reason for each of your answers.

i $\dfrac{X_1 + X_5 + X_n}{3}$

ii $\dfrac{\sum X^2}{n} - \mu$. (4)

2 The time people take to complete a simple puzzle is modelled by the continuous random variable X which is uniformly distributed over $1 \leqslant x \leqslant 5$.

a Write down the probability distribution function f(x). (1)

Find

b E(X), (1)

c Var(X), (2)

d P($1.5 \leqslant X \leqslant 4.5$). (2)

Given that John has already spent 3 minutes on the puzzle find the probability that John will complete the puzzle during the next minute. (2)

3 The probability of a mug produced at a pottery being faulty is 0.25.
Find the probability that in a random sample of 6 mugs

a exactly 1 mug is faulty, (3)

b more than 2 mugs are faulty. (2)

The mugs are sold in sets of six. Zafran buys 10 sets.

c Find the probability that exactly 3 of the sets contain more than 2 faulty mugs. (3)

4 **a** State two conditions under which a Poisson distribution is a suitable model to use in statistical work. (2)

A van hire company has 5 vans which they rent out by the day. Assuming that the number of vans hired out per day follows a Poisson distribution with mean 3, calculate, for a period of 100 days, the expected number of days when

b no vans will be hired, (3)

c the demand for van hire is not satisfied, (4)

d exactly 3 vans are hired. (3)

5 A manufacturer of coloured drawing pins introduces a purple pin that is to make up 15% of the total production. The pins are sold in boxes of 20.

a Find the critical region for a two-tailed test of the hypothesis that the probability that a pin chosen at random is purple is 0.15. The probability in either tail should be as close to 2.5% as possible. (3)

b Write down the actual significance level of the test above. (2)

A teacher buys a box of pins and discovers that it contains only 1 purple pin.

c Test, at the 5% level of significance, whether or not there is evidence that the probability of a pin chosen at random being purple is less than 0.15. (6)

6 The probability that a man is over 1.8 m tall is 0.2. A random sample of 50 men is taken and each man is measured and his height recorded.

a Find the probability that the number of men in the sample over 1.8 m tall is between 7 and 10 (inclusive) using

i a Poisson approximation,

ii a normal approximation. (10)

b By finding the actual value of the probability state which is the better apporoximation. (3)

7 A continuous random variable X has a probability density function given by

$$f(x) = \begin{cases} 1 - kx, & 1 \leqslant x \leqslant 3, \\ 0, & \text{otherwise.} \end{cases}$$

where k is a positive constant.

a Sketch f(x). (3)

b Show that $k = \frac{1}{4}$. (2)

c Calculate E(X). (3)

d Define fully the cumulative distribution function F(x). (4)

e Find the median of X. (3)

f Comment on the skewness of the distribution of X. (1)

Appendix

The normal distribution function

$\Phi(z) = P(Z < z)$

The function tabulated below is $\Phi(z)$, defined as $\Phi(z) = 1\backslash\sqrt{2\pi}\int_{-\infty}^{Z} e^{-\frac{1}{2}t^2}\,dt$.

z	$\Phi(z)$	z	$\Phi(z)$	z	$\Phi(z)$	z	$\Phi(z)$	z	$\Phi(z)$
0.00	0.5000	0.50	0.6915	1.00	0.8413	1.50	0.9332	2.00	0.9772
0.01	0.5040	0.51	0.6950	1.01	0.8438	1.51	0.9345	2.02	0.9783
0.02	0.5080	0.52	0.6985	1.02	0.8461	1.52	0.9357	2.04	0.9793
0.03	0.5120	0.53	0.7019	1.03	0.8485	1.53	0.9370	2.06	0.9803
0.04	0.5160	0.54	0.7054	1.04	0.8508	1.54	0.9382	2.08	0.9812
0.05	0.5199	0.55	0.7088	1.05	0.8531	1.55	0.9394	2.10	0.9821
0.06	0.5239	0.56	0.7123	1.06	0.8554	1.56	0.9406	2.12	0.9830
0.07	0.5279	0.57	0.7157	1.07	0.8577	1.57	0.9418	2.14	0.9838
0.08	0.5319	0.58	0.7190	1.08	0.8599	1.58	0.9429	2.16	0.9846
0.09	0.5359	0.59	0.7224	1.09	0.8621	1.59	0.9441	2.18	0.9854
0.10	0.5398	0.60	0.7257	1.10	0.8643	1.60	0.9452	2.20	0.9861
0.11	0.5438	0.61	0.7291	1.11	0.8665	1.61	0.9463	2.22	0.9868
0.12	0.5478	0.62	0.7324	1.12	0.8686	1.62	0.9474	2.24	0.9875
0.13	0.5517	0.63	0.7357	1.13	0.8708	1.63	0.9484	2.26	0.9881
0.14	0.5557	0.64	0.7389	1.14	0.8729	1.64	0.9495	2.28	0.9887
0.15	0.5596	0.65	0.7422	1.15	0.8749	1.65	0.9505	2.30	0.9893
0.16	0.5636	0.66	0.7454	1.16	0.8770	1.66	0.9515	2.32	0.9898
0.17	0.5675	0.67	0.7486	1.17	0.8790	1.67	0.9525	2.34	0.9904
0.18	0.5714	0.68	0.7517	1.18	0.8810	1.68	0.9535	2.36	0.9909
0.19	0.5753	0.69	0.7549	1.19	0.8830	1.69	0.9545	2.38	0.9913
0.20	0.5793	0.70	0.7580	1.20	0.8849	1.70	0.9554	2.40	0.9918
0.21	0.5832	0.71	0.7611	1.21	0.8869	1.71	0.9564	2.42	0.9922
0.22	0.5871	0.72	0.7642	1.22	0.8888	1.72	0.9573	2.44	0.9927
0.23	0.5910	0.73	0.7673	1.23	0.8907	1.73	0.9582	2.46	0.9931
0.24	0.5948	0.74	0.7704	1.24	0.8925	1.74	0.9591	2.48	0.9934
0.25	0.5987	0.75	0.7734	1.25	0.8944	1.75	0.9599	2.50	0.9938
0.26	0.6026	0.76	0.7764	1.26	0.8962	1.76	0.9608	2.55	0.9946
0.27	0.6064	0.77	0.7794	1.27	0.8980	1.77	0.9616	2.60	0.9953
0.28	0.6103	0.78	0.7823	1.28	0.8997	1.78	0.9625	2.65	0.9960
0.29	0.6141	0.79	0.7852	1.29	0.9015	1.79	0.9633	2.70	0.9965
0.30	0.6179	0.80	0.7881	1.30	0.9032	1.80	0.9641	2.75	0.9970
0.31	0.6217	0.81	0.7910	1.31	0.9049	1.81	0.9649	2.80	0.9974
0.32	0.6255	0.82	0.7939	1.32	0.9066	1.82	0.9656	2.85	0.9978
0.33	0.6293	0.83	0.7967	1.33	0.9082	1.83	0.9664	2.90	0.9981
0.34	0.6331	0.84	0.7995	1.34	0.9099	1.84	0.9671	2.95	0.9984
0.35	0.6368	0.85	0.8023	1.35	0.9115	1.85	0.9678	3.00	0.9987
0.36	0.6406	0.86	0.8051	1.36	0.9131	1.86	0.9686	3.05	0.9989
0.37	0.6443	0.87	0.8078	1.37	0.9147	1.87	0.9693	3.10	0.9990
0.38	0.6480	0.88	0.8106	1.38	0.9162	1.88	0.9699	3.15	0.9992
0.39	0.6517	0.89	0.8133	1.39	0.9177	1.89	0.9706	3.20	0.9993
0.40	0.6554	0.90	0.8159	1.40	0.9192	1.90	0.9713	3.25	0.9994
0.41	0.6591	0.91	0.8186	1.41	0.9207	1.91	0.9719	3.30	0.9995
0.42	0.6628	0.92	0.8212	1.42	0.9222	1.92	0.9726	3.35	0.9996
0.43	0.6664	0.93	0.8238	1.43	0.9236	1.93	0.9732	3.40	0.9997
0.44	0.6700	0.94	0.8264	1.44	0.9251	1.94	0.9738	3.50	0.9998
0.45	0.6736	0.95	0.8289	1.45	0.9265	1.95	0.9744	3.60	0.9998
0.46	0.6772	0.96	0.8315	1.46	0.9279	1.96	0.9750	3.70	0.9999
0.47	0.6808	0.97	0.8340	1.47	0.9292	1.97	0.9756	3.80	0.9999
0.48	0.6844	0.98	0.8365	1.48	0.9306	1.98	0.9761	3.90	1.0000
0.49	0.6879	0.99	0.8389	1.49	0.9319	1.99	0.9767	4.00	1.0000
0.50	0.6915	1.00	0.8413	1.50	0.9332	2.00	0.9772		

Percentage points of the normal distribution

The values z in the table are those which a random variable $Z \sim N(0,1)$ exceeds with probability p; that is, $P(Z > z) = p$.

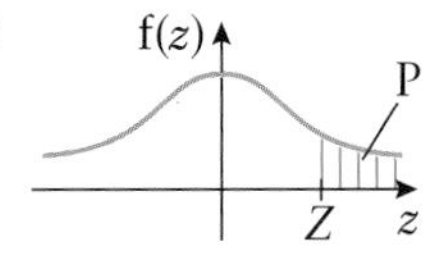

p	z	p	z
0.5000	0.0000	0.0500	1.6449
0.4000	0.2533	0.0250	1.9600
0.3000	0.5244	0.0100	2.3263
0.2000	0.8416	0.0050	2.5758
0.1500	1.0364	0.0010	3.0902
0.1000	1.2816	0.0005	3.2905

Statistics S2

The following formulae and tables are the ones in the Edexcel formula booklet under S2.

Discrete distributions

Standard discrete distributions:

Distribution of X	$P(X = x)$	Mean	Variance
Binomial $B(n, p)$	$\binom{n}{x} p^x (1-p)^{n-x}$	np	$np(1-p)$
Poisson $Po(\lambda)$	$e^{-\lambda}\frac{\lambda^x}{x!}$	λ	λ

Continuous distributions

For a continuous random variable X having probability density function f

Expectation (mean): $E(X) = \mu = \int x\,f(x)\,dx$

Variance: $Var(X) = \sigma^2 = \int (x-\mu)^2 f(x)\,dx = \int x^2 f(x)\,dx - \mu^2$

For a function $g(X)$: $E(g(X)) = \int g(x) f(x)\,dx$

Cumulative distribution function: $F(x_0) = P(X \leqslant x_0) = \int_{-\infty}^{x_0} f(t)\,dt$

Standard continuous distribution:

Distribution of X	P.D.F.	Mean	Variance
Uniform (Rectangular) on $[a, b]$	$\frac{1}{b-a}$	$\frac{1}{2}(a+b)$	$\frac{1}{12}(b-a)^2$

Binomial cumulative distribution function

The tabulated value is $P(X \leqslant x)$, where X has a binomial distribution with index n and parameter p.

$p =$	0.05	0.10	0.15	0.20	0.25	0.30	0.35	0.40	0.45	0.50
$n = 5, x = 0$	0.7738	0.5905	0.4437	0.3277	0.2373	0.1681	0.1160	0.0778	0.0503	0.0312
1	0.9774	0.9185	0.8352	0.7373	0.6328	0.5282	0.4284	0.3370	0.2562	0.1875
2	0.9988	0.9914	0.9734	0.9421	0.8965	0.8369	0.7648	0.6826	0.5931	0.5000
3	1.0000	0.9995	0.9978	0.9933	0.9844	0.9692	0.9460	0.9130	0.8688	0.8125
4	1.0000	1.0000	0.9999	0.9997	0.9990	0.9976	0.9947	0.9898	0.9815	0.9688
$n = 6, x = 0$	0.7351	0.5314	0.3771	0.2621	0.1780	0.1176	0.0754	0.0467	0.0277	0.0156
1	0.9672	0.8857	0.7765	0.6554	0.5339	0.4202	0.3191	0.2333	0.1636	0.1094
2	0.9978	0.9842	0.9527	0.9011	0.8306	0.7443	0.6471	0.5443	0.4415	0.3438
3	0.9999	0.9987	0.9941	0.9830	0.9624	0.9295	0.8826	0.8208	0.7447	0.6563
4	1.0000	0.9999	0.9996	0.9984	0.9954	0.9891	0.9777	0.9590	0.9308	0.8906
5	1.0000	1.0000	1.0000	0.9999	0.9998	0.9993	0.9982	0.9959	0.9917	0.9844
$n = 7, x = 0$	0.6983	0.4783	0.3206	0.2097	0.1335	0.0824	0.0490	0.0280	0.0152	0.0078
1	0.9556	0.8503	0.7166	0.5767	0.4449	0.3294	0.2338	0.1586	0.1024	0.0625
2	0.9962	0.9743	0.9262	0.8520	0.7564	0.6471	0.5323	0.4199	0.3164	0.2266
3	0.9998	0.9973	0.9879	0.9667	0.9294	0.8740	0.8002	0.7102	0.6083	0.5000
4	1.0000	0.9998	0.9988	0.9953	0.9871	0.9712	0.9444	0.9037	0.8471	0.7734
5	1.0000	1.0000	0.9999	0.9996	0.9987	0.9962	0.9910	0.9812	0.9643	0.9375
6	1.0000	1.0000	1.0000	1.0000	0.9999	0.9998	0.9994	0.9984	0.9963	0.9922
$n = 8, x = 0$	0.6634	0.4305	0.2725	0.1678	0.1001	0.0576	0.0319	0.0168	0.0084	0.0039
1	0.9428	0.8131	0.6572	0.5033	0.3671	0.2553	0.1691	0.1064	0.0632	0.0352
2	0.9942	0.9619	0.8948	0.7969	0.6785	0.5518	0.4278	0.3154	0.2201	0.1445
3	0.9996	0.9950	0.9786	0.9437	0.8862	0.8059	0.7064	0.5941	0.4770	0.3633
4	1.0000	0.9996	0.9971	0.9896	0.9727	0.9420	0.8939	0.8263	0.7396	0.6367
5	1.0000	1.0000	0.9998	0.9988	0.9958	0.9887	0.9747	0.9502	0.9115	0.8555
6	1.0000	1.0000	1.0000	0.9999	0.9996	0.9987	0.9964	0.9915	0.9819	0.9648
7	1.0000	1.0000	1.0000	1.0000	1.0000	0.9999	0.9998	0.9993	0.9983	0.9961
$n = 9, x = 0$	0.6302	0.3874	0.2316	0.1342	0.0751	0.0404	0.0207	0.0101	0.0046	0.0020
1	0.9288	0.7748	0.5995	0.4362	0.3003	0.1960	0.1211	0.0705	0.0385	0.0195
2	0.9916	0.9470	0.8591	0.7382	0.6007	0.4628	0.3373	0.2318	0.1495	0.0898
3	0.9994	0.9917	0.9661	0.9144	0.8343	0.7297	0.6089	0.4826	0.3614	0.2539
4	1.0000	0.9991	0.9944	0.9804	0.9511	0.9012	0.8283	0.7334	0.6214	0.5000
5	1.0000	0.9999	0.9994	0.9969	0.9900	0.9747	0.9464	0.9006	0.8342	0.7461
6	1.0000	1.0000	1.0000	0.9997	0.9987	0.9957	0.9888	0.9750	0.9502	0.9102
7	1.0000	1.0000	1.0000	1.0000	0.9999	0.9996	0.9986	0.9962	0.9909	0.9805
8	1.0000	1.0000	1.0000	1.0000	1.0000	1.0000	0.9999	0.9997	0.9992	0.9980
$n = 10, x = 0$	0.5987	0.3487	0.1969	0.1074	0.0563	0.0282	0.0135	0.0060	0.0025	0.0010
1	0.9139	0.7361	0.5443	0.3758	0.2440	0.1493	0.0860	0.0464	0.0233	0.0107
2	0.9885	0.9298	0.8202	0.6778	0.5256	0.3828	0.2616	0.1673	0.0996	0.0547
3	0.9990	0.9872	0.9500	0.8791	0.7759	0.6496	0.5138	0.3823	0.2660	0.1719
4	0.9999	0.9984	0.9901	0.9672	0.9219	0.8497	0.7515	0.6331	0.5044	0.3770
5	1.0000	0.9999	0.9986	0.9936	0.9803	0.9527	0.9051	0.8338	0.7384	0.6230
6	1.0000	1.0000	0.9999	0.9991	0.9965	0.9894	0.9740	0.9452	0.8980	0.8281
7	1.0000	1.0000	1.0000	0.9999	0.9996	0.9984	0.9952	0.9877	0.9726	0.9453
8	1.0000	1.0000	1.0000	1.0000	1.0000	0.9999	0.9995	0.9983	0.9955	0.9893
9	1.0000	1.0000	1.0000	1.0000	1.0000	1.0000	1.0000	0.9999	0.9997	0.9990

$p =$	0.05	0.10	0.15	0.20	0.25	0.30	0.35	0.40	0.45	0.50
$n = 12, x = 0$	0.5404	0.2824	0.1422	0.0687	0.0317	0.0138	0.0057	0.0022	0.0008	0.0002
1	0.8816	0.6590	0.4435	0.2749	0.1584	0.0850	0.0424	0.0196	0.0083	0.0032
2	0.9804	0.8891	0.7358	0.5583	0.3907	0.2528	0.1513	0.0834	0.0421	0.0193
3	0.9978	0.9744	0.9078	0.7946	0.6488	0.4925	0.3467	0.2253	0.1345	0.0730
4	0.9998	0.9957	0.9761	0.9274	0.8424	0.7237	0.5833	0.4382	0.3044	0.1938
5	1.0000	0.9995	0.9954	0.9806	0.9456	0.8822	0.7873	0.6652	0.5269	0.3872
6	1.0000	0.9999	0.9993	0.9961	0.9857	0.9614	0.9154	0.8418	0.7393	0.6128
7	1.0000	1.0000	0.9999	0.9994	0.9972	0.9905	0.9745	0.9427	0.8883	0.8062
8	1.0000	1.0000	1.0000	0.9999	0.9996	0.9983	0.9944	0.9847	0.9644	0.9270
9	1.0000	1.0000	1.0000	1.0000	1.0000	0.9998	0.9992	0.9972	0.9921	0.9807
10	1.0000	1.0000	1.0000	1.0000	1.0000	1.0000	0.9999	0.9997	0.9989	0.9968
11	1.0000	1.0000	1.0000	1.0000	1.0000	1.0000	1.0000	1.0000	0.9999	0.9998
$n = 15, x = 0$	0.4633	0.2059	0.0874	0.0352	0.0134	0.0047	0.0016	0.0005	0.0001	0.0000
1	0.8290	0.5490	0.3186	0.1671	0.0802	0.0353	0.0142	0.0052	0.0017	0.0005
2	0.9638	0.8159	0.6042	0.3980	0.2361	0.1268	0.0617	0.0271	0.0107	0.0037
3	0.9945	0.9444	0.8227	0.6482	0.4613	0.2969	0.1727	0.0905	0.0424	0.0176
4	0.9994	0.9873	0.9383	0.8358	0.6865	0.5155	0.3519	0.2173	0.1204	0.0592
5	0.9999	0.9978	0.9832	0.9389	0.8516	0.7216	0.5643	0.4032	0.2608	0.1509
6	1.0000	0.9997	0.9964	0.9819	0.9434	0.8689	0.7548	0.6098	0.4522	0.3036
7	1.0000	1.0000	0.9994	0.9958	0.9827	0.9500	0.8868	0.7869	0.6535	0.5000
8	1.0000	1.0000	0.9999	0.9992	0.9958	0.9848	0.9578	0.9050	0.8182	0.6964
9	1.0000	1.0000	1.0000	0.9999	0.9992	0.9963	0.9876	0.9662	0.9231	0.8491
10	1.0000	1.0000	1.0000	1.0000	0.9999	0.9993	0.9972	0.9907	0.9745	0.9408
11	1.0000	1.0000	1.0000	1.0000	1.0000	0.9999	0.9995	0.9981	0.9937	0.9824
12	1.0000	1.0000	1.0000	1.0000	1.0000	1.0000	0.9999	0.9997	0.9989	0.9963
13	1.0000	1.0000	1.0000	1.0000	1.0000	1.0000	1.0000	1.0000	0.9999	0.9995
14	1.0000	1.0000	1.0000	1.0000	1.0000	1.0000	1.0000	1.0000	1.0000	1.0000
$n = 20, x = 0$	0.3585	0.1216	0.0388	0.0115	0.0032	0.0008	0.0002	0.0000	0.0000	0.0000
1	0.7358	0.3917	0.1756	0.0692	0.0243	0.0076	0.0021	0.0005	0.0001	0.0000
2	0.9245	0.6769	0.4049	0.2061	0.0913	0.0355	0.0121	0.0036	0.0009	0.0002
3	0.9841	0.8670	0.6477	0.4114	0.2252	0.1071	0.0444	0.0160	0.0049	0.0013
4	0.9974	0.9568	0.8298	0.6296	0.4148	0.2375	0.1182	0.0510	0.0189	0.0059
5	0.9997	0.9887	0.9327	0.8042	0.6172	0.4164	0.2454	0.1256	0.0553	0.0207
6	1.0000	0.9976	0.9781	0.9133	0.7858	0.6080	0.4166	0.2500	0.1299	0.0577
7	1.0000	0.9996	0.9941	0.9679	0.8982	0.7723	0.6010	0.4159	0.2520	0.1316
8	1.0000	0.9999	0.9987	0.9900	0.9591	0.8867	0.7624	0.5956	0.4143	0.2517
9	1.0000	1.0000	0.9998	0.9974	0.9861	0.9520	0.8782	0.7553	0.5914	0.4119
10	1.0000	1.0000	1.0000	0.9994	0.9961	0.9829	0.9468	0.8725	0.7507	0.5881
11	1.0000	1.0000	1.0000	0.9999	0.9991	0.9949	0.9804	0.9435	0.8692	0.7483
12	1.0000	1.0000	1.0000	1.0000	0.9998	0.9987	0.9940	0.9790	0.9420	0.8684
13	1.0000	1.0000	1.0000	1.0000	1.0000	0.9997	0.9985	0.9935	0.9786	0.9423
14	1.0000	1.0000	1.0000	1.0000	1.0000	1.0000	0.9997	0.9984	0.9936	0.9793
15	1.0000	1.0000	1.0000	1.0000	1.0000	1.0000	1.0000	0.9997	0.9985	0.9941
16	1.0000	1.0000	1.0000	1.0000	1.0000	1.0000	1.0000	1.0000	0.9997	0.9987
17	1.0000	1.0000	1.0000	1.0000	1.0000	1.0000	1.0000	1.0000	1.0000	0.9998
18	1.0000	1.0000	1.0000	1.0000	1.0000	1.0000	1.0000	1.0000	1.0000	1.0000

$p =$	0.05	0.10	0.15	0.20	0.25	0.30	0.35	0.40	0.45	0.50
$n = 25, x = 0$	0.2774	0.0718	0.0172	0.0038	0.0008	0.0001	0.0000	0.0000	0.0000	0.0000
1	0.6424	0.2712	0.0931	0.0274	0.0070	0.0016	0.0003	0.0001	0.0000	0.0000
2	0.8729	0.5371	0.2537	0.0982	0.0321	0.0090	0.0021	0.0004	0.0001	0.0000
3	0.9659	0.7636	0.4711	0.2340	0.0962	0.0332	0.0097	0.0024	0.0005	0.0001
4	0.9928	0.9020	0.6821	0.4207	0.2137	0.0905	0.0320	0.0095	0.0023	0.0005
5	0.9988	0.9666	0.8385	0.6167	0.3783	0.1935	0.0826	0.0294	0.0086	0.0020
6	0.9998	0.9905	0.9305	0.7800	0.5611	0.3407	0.1734	0.0736	0.0258	0.0073
7	1.0000	0.9977	0.9745	0.8909	0.7265	0.5118	0.3061	0.1536	0.0639	0.0216
8	1.0000	0.9995	0.9920	0.9532	0.8506	0.6769	0.4668	0.2735	0.1340	0.0539
9	1.0000	0.9999	0.9979	0.9827	0.9287	0.8106	0.6303	0.4246	0.2424	0.1148
10	1.0000	1.0000	0.9995	0.9944	0.9703	0.9022	0.7712	0.5858	0.3843	0.2122
11	1.0000	1.0000	0.9999	0.9985	0.9893	0.9558	0.8746	0.7323	0.5426	0.3450
12	1.0000	1.0000	1.0000	0.9996	0.9966	0.9825	0.9396	0.8462	0.6937	0.5000
13	1.0000	1.0000	1.0000	0.9999	0.9991	0.9940	0.9745	0.9222	0.8173	0.6550
14	1.0000	1.0000	1.0000	1.0000	0.9998	0.9982	0.9907	0.9656	0.9040	0.7878
15	1.0000	1.0000	1.0000	1.0000	1.0000	0.9995	0.9971	0.9868	0.9560	0.8852
16	1.0000	1.0000	1.0000	1.0000	1.0000	0.9999	0.9992	0.9957	0.9826	0.9461
17	1.0000	1.0000	1.0000	1.0000	1.0000	1.0000	0.9998	0.9988	0.9942	0.9784
18	1.0000	1.0000	1.0000	1.0000	1.0000	1.0000	1.0000	0.9997	0.9984	0.9927
19	1.0000	1.0000	1.0000	1.0000	1.0000	1.0000	1.0000	0.9999	0.9996	0.9980
20	1.0000	1.0000	1.0000	1.0000	1.0000	1.0000	1.0000	1.0000	0.9999	0.9995
21	1.0000	1.0000	1.0000	1.0000	1.0000	1.0000	1.0000	1.0000	1.0000	0.9999
22	1.0000	1.0000	1.0000	1.0000	1.0000	1.0000	1.0000	1.0000	1.0000	1.0000
$n = 30, x = 0$	0.2146	0.0424	0.0076	0.0012	0.0002	0.0000	0.0000	0.0000	0.0000	0.0000
1	0.5535	0.1837	0.0480	0.0105	0.0020	0.0003	0.0000	0.0000	0.0000	0.0000
2	0.8122	0.4114	0.1514	0.0442	0.0106	0.0021	0.0003	0.0000	0.0000	0.0000
3	0.9392	0.6474	0.3217	0.1227	0.0374	0.0093	0.0019	0.0003	0.0000	0.0000
4	0.9844	0.8245	0.5245	0.2552	0.0979	0.0302	0.0075	0.0015	0.0002	0.0000
5	0.9967	0.9268	0.7106	0.4275	0.2026	0.0766	0.0233	0.0057	0.0011	0.0002
6	0.9994	0.9742	0.8474	0.6070	0.3481	0.1595	0.0586	0.0172	0.0040	0.0007
7	0.9999	0.9922	0.9302	0.7608	0.5143	0.2814	0.1238	0.0435	0.0121	0.0026
8	1.0000	0.9980	0.9722	0.8713	0.6736	0.4315	0.2247	0.0940	0.0312	0.0081
9	1.0000	0.9995	0.9903	0.9389	0.8034	0.5888	0.3575	0.1763	0.0694	0.0214
10	1.0000	0.9999	0.9971	0.9744	0.8943	0.7304	0.5078	0.2915	0.1350	0.0494
11	1.0000	1.0000	0.9992	0.9905	0.9493	0.8407	0.6548	0.4311	0.2327	0.1002
12	1.0000	1.0000	0.9998	0.9969	0.9784	0.9155	0.7802	0.5785	0.3592	0.1808
13	1.0000	1.0000	1.0000	0.9991	0.9918	0.9599	0.8737	0.7145	0.5025	0.2923
14	1.0000	1.0000	1.0000	0.9998	0.9973	0.9831	0.9348	0.8246	0.6448	0.4278
15	1.0000	1.0000	1.0000	0.9999	0.9992	0.9936	0.9699	0.9029	0.7691	0.5722
16	1.0000	1.0000	1.0000	1.0000	0.9998	0.9979	0.9876	0.9519	0.8644	0.7077
17	1.0000	1.0000	1.0000	1.0000	0.9999	0.9994	0.9955	0.9788	0.9286	0.8192
18	1.0000	1.0000	1.0000	1.0000	1.0000	0.9998	0.9986	0.9917	0.9666	0.8998
19	1.0000	1.0000	1.0000	1.0000	1.0000	1.0000	0.9996	0.9971	0.9862	0.9506
20	1.0000	1.0000	1.0000	1.0000	1.0000	1.0000	0.9999	0.9991	0.9950	0.9786
21	1.0000	1.0000	1.0000	1.0000	1.0000	1.0000	1.0000	0.9998	0.9984	0.9919
22	1.0000	1.0000	1.0000	1.0000	1.0000	1.0000	1.0000	1.0000	0.9996	0.9974
23	1.0000	1.0000	1.0000	1.0000	1.0000	1.0000	1.0000	1.0000	0.9999	0.9993
24	1.0000	1.0000	1.0000	1.0000	1.0000	1.0000	1.0000	1.0000	1.0000	0.9998
25	1.0000	1.0000	1.0000	1.0000	1.0000	1.0000	1.0000	1.0000	1.0000	1.0000

$p =$	0.05	0.10	0.15	0.20	0.25	0.30	0.35	0.40	0.45	0.50
$n = 40, x = 0$	0.1285	0.0148	0.0015	0.0001	0.0000	0.0000	0.0000	0.0000	0.0000	0.0000
1	0.3991	0.0805	0.0121	0.0015	0.0001	0.0000	0.0000	0.0000	0.0000	0.0000
2	0.6767	0.2228	0.0486	0.0079	0.0010	0.0001	0.0000	0.0000	0.0000	0.0000
3	0.8619	0.4231	0.1302	0.0285	0.0047	0.0006	0.0001	0.0000	0.0000	0.0000
4	0.9520	0.6290	0.2633	0.0759	0.0160	0.0026	0.0003	0.0000	0.0000	0.0000
5	0.9861	0.7937	0.4325	0.1613	0.0433	0.0086	0.0013	0.0001	0.0000	0.0000
6	0.9966	0.9005	0.6067	0.2859	0.0962	0.0238	0.0044	0.0006	0.0001	0.0000
7	0.9993	0.9581	0.7559	0.4371	0.1820	0.0553	0.0124	0.0021	0.0002	0.0000
8	0.9999	0.9845	0.8646	0.5931	0.2998	0.1110	0.0303	0.0061	0.0009	0.0001
9	1.0000	0.9949	0.9328	0.7318	0.4395	0.1959	0.0644	0.0156	0.0027	0.0003
10	1.0000	0.9985	0.9701	0.8392	0.5839	0.3087	0.1215	0.0352	0.0074	0.0011
11	1.0000	0.9996	0.9880	0.9125	0.7151	0.4406	0.2053	0.0709	0.0179	0.0032
12	1.0000	0.9999	0.9957	0.9568	0.8209	0.5772	0.3143	0.1285	0.0386	0.0083
13	1.0000	1.0000	0.9986	0.9806	0.8968	0.7032	0.4408	0.2112	0.0751	0.0192
14	1.0000	1.0000	0.9996	0.9921	0.9456	0.8074	0.5721	0.3174	0.1326	0.0403
15	1.0000	1.0000	0.9999	0.9971	0.9738	0.8849	0.6946	0.4402	0.2142	0.0769
16	1.0000	1.0000	1.0000	0.9990	0.9884	0.9367	0.7978	0.5681	0.3185	0.1341
17	1.0000	1.0000	1.0000	0.9997	0.9953	0.9680	0.8761	0.6885	0.4391	0.2148
18	1.0000	1.0000	1.0000	0.9999	0.9983	0.9852	0.9301	0.7911	0.5651	0.3179
19	1.0000	1.0000	1.0000	1.0000	0.9994	0.9937	0.9637	0.8702	0.6844	0.4373
20	1.0000	1.0000	1.0000	1.0000	0.9998	0.9976	0.9827	0.9256	0.7870	0.5627
21	1.0000	1.0000	1.0000	1.0000	1.0000	0.9991	0.9925	0.9608	0.8669	0.6821
22	1.0000	1.0000	1.0000	1.0000	1.0000	0.9997	0.9970	0.9811	0.9233	0.7852
23	1.0000	1.0000	1.0000	1.0000	1.0000	0.9999	0.9989	0.9917	0.9595	0.8659
24	1.0000	1.0000	1.0000	1.0000	1.0000	1.0000	0.9996	0.9966	0.9804	0.9231
25	1.0000	1.0000	1.0000	1.0000	1.0000	1.0000	0.9999	0.9988	0.9914	0.9597
26	1.0000	1.0000	1.0000	1.0000	1.0000	1.0000	1.0000	0.9996	0.9966	0.9808
27	1.0000	1.0000	1.0000	1.0000	1.0000	1.0000	1.0000	0.9999	0.9988	0.9917
28	1.0000	1.0000	1.0000	1.0000	1.0000	1.0000	1.0000	1.0000	0.9996	0.9968
29	1.0000	1.0000	1.0000	1.0000	1.0000	1.0000	1.0000	1.0000	0.9999	0.9989
30	1.0000	1.0000	1.0000	1.0000	1.0000	1.0000	1.0000	1.0000	1.0000	0.9997
31	1.0000	1.0000	1.0000	1.0000	1.0000	1.0000	1.0000	1.0000	1.0000	0.9999
32	1.0000	1.0000	1.0000	1.0000	1.0000	1.0000	1.0000	1.0000	1.0000	1.0000

$p =$	0.05	0.10	0.15	0.20	0.25	0.30	0.35	0.40	0.45	0.50
$n = 50, x = 0$	0.0769	0.0052	0.0003	0.0000	0.0000	0.0000	0.0000	0.0000	0.0000	0.0000
1	0.2794	0.0338	0.0029	0.0002	0.0000	0.0000	0.0000	0.0000	0.0000	0.0000
2	0.5405	0.1117	0.0142	0.0013	0.0001	0.0000	0.0000	0.0000	0.0000	0.0000
3	0.7604	0.2503	0.0460	0.0057	0.0005	0.0000	0.0000	0.0000	0.0000	0.0000
4	0.8964	0.4312	0.1121	0.0185	0.0021	0.0002	0.0000	0.0000	0.0000	0.0000
5	0.9622	0.6161	0.2194	0.0480	0.0070	0.0007	0.0001	0.0000	0.0000	0.0000
6	0.9882	0.7702	0.3613	0.1034	0.0194	0.0025	0.0002	0.0000	0.0000	0.0000
7	0.9968	0.8779	0.5188	0.1904	0.0453	0.0073	0.0008	0.0001	0.0000	0.0000
8	0.9992	0.9421	0.6681	0.3073	0.0916	0.0183	0.0025	0.0002	0.0000	0.0000
9	0.9998	0.9755	0.7911	0.4437	0.1637	0.0402	0.0067	0.0008	0.0001	0.0000
10	1.0000	0.9906	0.8801	0.5836	0.2622	0.0789	0.0160	0.0022	0.0002	0.0000
11	1.0000	0.9968	0.9372	0.7107	0.3816	0.1390	0.0342	0.0057	0.0006	0.0000
12	1.0000	0.9990	0.9699	0.8139	0.5110	0.2229	0.0661	0.0133	0.0018	0.0002
13	1.0000	0.9997	0.9868	0.8894	0.6370	0.3279	0.1163	0.0280	0.0045	0.0005
14	1.0000	0.9999	0.9947	0.9393	0.7481	0.4468	0.1878	0.0540	0.0104	0.0013
15	1.0000	1.0000	0.9981	0.9692	0.8369	0.5692	0.2801	0.0955	0.0220	0.0033
16	1.0000	1.0000	0.9993	0.9856	0.9017	0.6839	0.3889	0.1561	0.0427	0.0077
17	1.0000	1.0000	0.9998	0.9937	0.9449	0.7822	0.5060	0.2369	0.0765	0.0164
18	1.0000	1.0000	0.9999	0.9975	0.9713	0.8594	0.6216	0.3356	0.1273	0.0325
19	1.0000	1.0000	1.0000	0.9991	0.9861	0.9152	0.7264	0.4465	0.1974	0.0595
20	1.0000	1.0000	1.0000	0.9997	0.9937	0.9522	0.8139	0.5610	0.2862	0.1013
21	1.0000	1.0000	1.0000	0.9999	0.9974	0.9749	0.8813	0.6701	0.3900	0.1611
22	1.0000	1.0000	1.0000	1.0000	0.9990	0.9877	0.9290	0.7660	0.5019	0.2399
23	1.0000	1.0000	1.0000	1.0000	0.9996	0.9944	0.9604	0.8438	0.6134	0.3359
24	1.0000	1.0000	1.0000	1.0000	0.9999	0.9976	0.9793	0.9022	0.7160	0.4439
25	1.0000	1.0000	1.0000	1.0000	1.0000	0.9991	0.9900	0.9427	0.8034	0.5561
26	1.0000	1.0000	1.0000	1.0000	1.0000	0.9997	0.9955	0.9686	0.8721	0.6641
27	1.0000	1.0000	1.0000	1.0000	1.0000	0.9999	0.9981	0.9840	0.9220	0.7601
28	1.0000	1.0000	1.0000	1.0000	1.0000	1.0000	0.9993	0.9924	0.9556	0.8389
29	1.0000	1.0000	1.0000	1.0000	1.0000	1.0000	0.9997	0.9966	0.9765	0.8987
30	1.0000	1.0000	1.0000	1.0000	1.0000	1.0000	0.9999	0.9986	0.9884	0.9405
31	1.0000	1.0000	1.0000	1.0000	1.0000	1.0000	1.0000	0.9995	0.9947	0.9675
32	1.0000	1.0000	1.0000	1.0000	1.0000	1.0000	1.0000	0.9998	0.9978	0.9836
33	1.0000	1.0000	1.0000	1.0000	1.0000	1.0000	1.0000	0.9999	0.9991	0.9923
34	1.0000	1.0000	1.0000	1.0000	1.0000	1.0000	1.0000	1.0000	0.9997	0.9967
35	1.0000	1.0000	1.0000	1.0000	1.0000	1.0000	1.0000	1.0000	0.9999	0.9987
36	1.0000	1.0000	1.0000	1.0000	1.0000	1.0000	1.0000	1.0000	1.0000	0.9995
37	1.0000	1.0000	1.0000	1.0000	1.0000	1.0000	1.0000	1.0000	1.0000	0.9998
38	1.0000	1.0000	1.0000	1.0000	1.0000	1.0000	1.0000	1.0000	1.0000	1.0000

POISSON CUMULATIVE DISTRIBUTION FUNCTION

The tabulated value is $P(X \leq x)$, where X has a Poisson distribution with parameter λ.

$\lambda =$	0.5	1.0	1.5	2.0	2.5	3.0	3.5	4.0	4.5	5.0
$x = 0$	0.6065	0.3679	0.2231	0.1353	0.0821	0.0498	0.0302	0.0183	0.0111	0.0067
1	0.9098	0.7358	0.5578	0.4060	0.2873	0.1991	0.1359	0.0916	0.0611	0.0404
2	0.9856	0.9197	0.8088	0.6767	0.5438	0.4232	0.3208	0.2381	0.1736	0.1247
3	0.9982	0.9810	0.9344	0.8571	0.7576	0.6472	0.5366	0.4335	0.3423	0.2650
4	0.9998	0.9963	0.9814	0.9473	0.8912	0.8153	0.7254	0.6288	0.5321	0.4405
5	1.0000	0.9994	0.9955	0.9834	0.9580	0.9161	0.8576	0.7851	0.7029	0.6160
6	1.0000	0.9999	0.9991	0.9955	0.9858	0.9665	0.9347	0.8893	0.8311	0.7622
7	1.0000	1.0000	0.9998	0.9989	0.9958	0.9881	0.9733	0.9489	0.9134	0.8666
8	1.0000	1.0000	1.0000	0.9998	0.9989	0.9962	0.9901	0.9786	0.9597	0.9319
9	1.0000	1.0000	1.0000	1.0000	0.9997	0.9989	0.9967	0.9919	0.9829	0.9682
10	1.0000	1.0000	1.0000	1.0000	0.9999	0.9997	0.9990	0.9972	0.9933	0.9863
11	1.0000	1.0000	1.0000	1.0000	1.0000	0.9999	0.9997	0.9991	0.9976	0.9945
12	1.0000	1.0000	1.0000	1.0000	1.0000	1.0000	0.9999	0.9997	0.9992	0.9980
13	1.0000	1.0000	1.0000	1.0000	1.0000	1.0000	1.0000	0.9999	0.9997	0.9993
14	1.0000	1.0000	1.0000	1.0000	1.0000	1.0000	1.0000	1.0000	0.9999	0.9998
15	1.0000	1.0000	1.0000	1.0000	1.0000	1.0000	1.0000	1.0000	1.0000	0.9999
16	1.0000	1.0000	1.0000	1.0000	1.0000	1.0000	1.0000	1.0000	1.0000	1.0000
17	1.0000	1.0000	1.0000	1.0000	1.0000	1.0000	1.0000	1.0000	1.0000	1.0000
18	1.0000	1.0000	1.0000	1.0000	1.0000	1.0000	1.0000	1.0000	1.0000	1.0000
19	1.0000	1.0000	1.0000	1.0000	1.0000	1.0000	1.0000	1.0000	1.0000	1.0000

$\lambda =$	5.5	6.0	6.5	7.0	7.5	8.0	8.5	9.0	9.5	10.0
$x = 0$	0.0041	0.0025	0.0015	0.0009	0.0006	0.0003	0.0002	0.0001	0.0001	0.0000
1	0.0266	0.0174	0.0113	0.0073	0.0047	0.0030	0.0019	0.0012	0.0008	0.0005
2	0.0884	0.0620	0.0430	0.0296	0.0203	0.0138	0.0093	0.0062	0.0042	0.0028
3	0.2017	0.1512	0.1118	0.0818	0.0591	0.0424	0.0301	0.0212	0.0149	0.0103
4	0.3575	0.2851	0.2237	0.1730	0.1321	0.0996	0.0744	0.0550	0.0403	0.0293
5	0.5289	0.4457	0.3690	0.3007	0.2414	0.1912	0.1496	0.1157	0.0885	0.0671
6	0.6860	0.6063	0.5265	0.4497	0.3782	0.3134	0.2562	0.2068	0.1649	0.1301
7	0.8095	0.7440	0.6728	0.5987	0.5246	0.4530	0.3856	0.3239	0.2687	0.2202
8	0.8944	0.8472	0.7916	0.7291	0.6620	0.5925	0.5231	0.4557	0.3918	0.3328
9	0.9462	0.9161	0.8774	0.8305	0.7764	0.7166	0.6530	0.5874	0.5218	0.4579
10	0.9747	0.9574	0.9332	0.9015	0.8622	0.8159	0.7634	0.7060	0.6453	0.5830
11	0.9890	0.9799	0.9661	0.9467	0.9208	0.8881	0.8487	0.8030	0.7520	0.6968
12	0.9955	0.9912	0.9840	0.9730	0.9573	0.9362	0.9091	0.8758	0.8364	0.7916
13	0.9983	0.9964	0.9929	0.9872	0.9784	0.9658	0.9486	0.9261	0.8981	0.8645
14	0.9994	0.9986	0.9970	0.9943	0.9897	0.9827	0.9726	0.9585	0.9400	0.9165
15	0.9998	0.9995	0.9988	0.9976	0.9954	0.9918	0.9862	0.9780	0.9665	0.9513
16	0.9999	0.9998	0.9996	0.9990	0.9980	0.9963	0.9934	0.9889	0.9823	0.9730
17	1.0000	0.9999	0.9998	0.9996	0.9992	0.9984	0.9970	0.9947	0.9911	0.9857
18	1.0000	1.0000	0.9999	0.9999	0.9997	0.9993	0.9987	0.9976	0.9957	0.9928
19	1.0000	1.0000	1.0000	1.0000	0.9999	0.9997	0.9995	0.9989	0.9980	0.9965
20	1.0000	1.0000	1.0000	1.0000	1.0000	0.9999	0.9998	0.9996	0.9991	0.9984
21	1.0000	1.0000	1.0000	1.0000	1.0000	1.0000	0.9999	0.9998	0.9996	0.9993
22	1.0000	1.0000	1.0000	1.0000	1.0000	1.0000	1.0000	0.9999	0.9999	0.9997

Answers

Exercise 1A

1 a 120 b 10 c 21
d 210 e 190

2 a $\frac{8}{429}$ or 0.0186 b $\frac{16}{143}$ or 0.112
c 0.000999 or $\frac{1}{1001}$

3 a 0.279 b 0.0781 c 0.000125

Exercise 1B

1 a 0.273 b 0.0683 c 0.195
2 a 0.132 b 0.356 c 0.00464
3 a 0.00670 b 0.214 c 0.00178
4 a 0.358 b 0.189
5 a $X \sim B(20, 0.01)$
$n = 20$
$p = 0.01$
Assume bolts being defective are independent of each other.
b $X \sim B(6, 0.52)$
$n = 6$
$p = 0.52$
Assume the lights operate independently and the time lights are on/off is constant.
c $X \sim B(30, \frac{1}{8})$
$n = 30$
$p = \frac{1}{8}$
Assume serves are independent and probability of an ace is constant.
6 a $X \sim B(14, 0.15)$ is OK if we assume the children in the class being Rh^- is independent from child to child (so no siblings/twins).
b This is not binomial since the number of tosses is not fixed.
The probability of a head at each toss is constant ($p = 0.5$) but there is no value of n.
c Assuming the colours of the cars are independent (which should be reasonable).
X = number of red cars out of 15
$X \sim B(15, 0.12)$
7 a $\frac{125}{1296}$ or 0.0965 b 0.155
8 a $\frac{16}{243}$ or 0.0658 b 0.307 or $\frac{224}{729}$

Exercise 1C

1 a 0.9804 b 0.7382 c 0.5638 d 0.3020
2 a 0.9468 b 0.5834 c 0.1272 d 0.5989
3 a 0.6844 b 0.6815 c 0.2068 d 0.3111
4 a 0.0278 b 0.5245 c 0.9423 d 0.2028
5 a 0.0039 b 0.9648 c 0.3633
6 a 0.2252 b 0.4613 c 0.7073
7 a $k = 13$ b $r = 28$
8 a $k = 1$ b $r = 9$ c 0.9801
9 a $X \sim B(10, 0.30)$
b 0.1503 c $s = 8$
10 a 0.2794 b 0.0378 c $d = 5$

Exercise 1D

1 $E(X) = 6$
$Var(X) = 5$
2 a $\frac{9}{4}$ or 2.25 b 0.6840
3 a $E(X) = 12$
$Var(X) = 7.2$
b 0.4022
4 a $n = 60$ b 1.69
5 $n = 7$
6 $p = \frac{1}{5}$ and $n = 225$

Mixed exercise 1E

1 a 0.0439 or $\frac{32}{729}$ b 0.273
2 a 0.0138 (3 s.f.) b 0.747 (3 d.p.)
3 a 0.987 0.983 b 10.9
4 a 1 There are n independent trials.
2 n is a fixed number.
3 The outcome of each trial is success or failure.
4 The probability of success at each trial is constant.
5 The outcome of any trial is independant of any other trial.
b 0.0861 c $n = 60$ d $n = 90$
5 a 0.879 b 0.773
6 a 0.000977 b 0.0547
7 a 0.0531 b 0.243
8 a $X \sim B(10, 0.15)$
b 0.0099 c 0.2759
9 a 0.8692 b 0.0728

Exercise 2A

1 a 0.117 b 0.900 c 0.0538
2 a 0.130 b 0.0224 c 0.923
3 a 0.247 b 0.295 c 0.290
4 a 0.138 b 0.412 c 0.126

Exercise 2B

1 a 0.2052 b 0.4562 c 0.9580 d 0.4142
2 a 0.1512 b 0.7149 c 0.1606 d 0.6820
3 a 0.1125 b 0.0611 c 0.4679 d 0.7700
4 a $a = 6$ b $b = 9$ c $c = 5$ d $d = 5$
5 a $a = 5$ b $b = 2$ c $c \geqslant 7$ d $d \geqslant 9$
6 a 0.00248 or 0.0025
b 0.1512 c 0.0174 d 0.0620
7 a 9 b 3 c 0.3473 d 0.2068
8 a 0.2231 b 0.8088

Exercise 2C

1 a 0.1730 b 0.0302

2 a 0.0235 (3 s.f.) b 0.0293
  c Assume that defects occur independently and at random in the cloth and defects occur at a constant rate.

3 a If the misprints occur independently and at random and at a constant average rate then this could be Poisson.
  b Yes because after 1 hour the pigs are probably dispersed fairly randomly and independently around the field.
  c No because after 1 minute the pigs will probably be clustered around the feeding trough and so will not be randomly and independently scattered.
  d No because the salt needs to diffuse so that it is randomly dissolved at a constant average rate throughout the contents of the bucket.
  e Yes, this may be Poisson provided that the runners are not in groups, since they need to pass the post independently and at random.

4 a 0.0668 b 0.0835

5 a 0.2510 b 0.5874

6 a $X \sim \text{Po}(2.5)$ b 0.2424
  c The sheep will no longer be randomly scattered.

7 a 0.5578 b 0.0656 c 0.0025

8 a 0.513 (3 s.f.) b $n = 6$

9 a 0.2971 b $0.0038 < 0.01$
  c They would need 10 boats.

10 a 0.8088 b 0.1847 c $n = 14$

Exercise 2D

1 a 0.9997 b 0.3134

2 a 0.218 b 0.430

3 0.58%

4 a 0.0174 b 0.1339

5 a 0.0235 (3 s.f.) b 0.8883

6 a 0.0337 (3 s.f.) b 0.0842 (3 s.f.)
  c 0.875 (3 s.f.)

7 a 0.0183 (Poisson table)
  b 0.0733 (Poisson table)
  c 0.2381 (Poisson table)

8 a 0.0000454 (3 s.f.) b 0.0671

9 a For large n and small p
  $\text{B}(n, p) \approx \text{Po}(np)$
  b 0.00992 c 0.3233

Exercise 2E

1 a If the outcomes occur:
  1 singly
  2 at a constant rate
  3 independently and at random
  then a Poisson distribution can be suitable.
  b 0.9502 c 0.423 d 0.702

2 a 0.879 (3 s.f.) b 0.773 (3 s.f.)
  c 0.7764 (Poisson tables) d $m = 7$

3 a 0.0177 b 0.2236 c 0.1251 d 0.0773

4 a 0.221 (3 d.p.) b 0.998 (3 d.p.)
  c 0.050 (3 d.p.) d 0.925 (3 d.p.)

5 a $X \sim \text{Po}(0.25)$ b 0.221

Mixed exercise 2F

1 a i 0.223 (3 d.p.) ii 0.442 (3 d.p.)
  b 0.0987 (3 s.f.)

2 Items occur in continuous space or time:
  1 singly
  2 at a constant rate
  3 independently of one another and at random.
  a 0.3032 b 0.0279 (3 s.f.)

3 a $\text{B}(n, p)$ can be approximated to $\text{Po}(np)$
  if n is large and p is small
  then mean $= np$
  b i 0.908 (3 d.p.) ii 0.00370 (3 s.f.)
  c 0.000264 (3 s.f.)

4 a Binomial b Poisson c Poisson
  d Binomial e Poisson f Poisson

5 a 10 b 0.0067

6 a $X \sim \text{Po}(0.25)$ b 0.0265 c 0.276 (3 s.f.)

7 a 0.2138 b 0.1378 c 0.0457

8 a $X \sim \text{Po}(0.6)$ b 0.549 (3 s.f.) c 0.0231 (3 s.f.)
  d Not suitable.
  The rate of telephone boxes will be different in cities and they are more likely to occur in clusters.

9 a 0.0756 b 0.925 (3 d.p.)
  c 0.065 d 0.4005 (4 d.p.)

10 a 0.181 (3 d.p.) b 0.0191

Exercise 3A

1 a There are negative values for f(x) when $x < 0$ so this is not a probability density function.
  b Area of $8\frac{2}{3}$ not equal to 1 therefore it is not a valid probability density function.
  c There are negative values for f(x) so this is not a probability density function.

2 $k = -\frac{3}{50}$

3 a

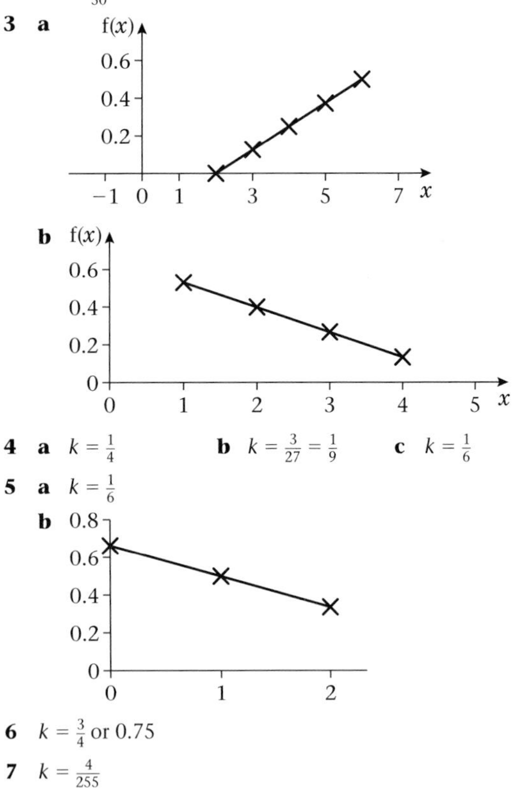

4 a $k = \frac{1}{4}$ b $k = \frac{3}{27} = \frac{1}{9}$ c $k = \frac{1}{6}$

5 a $k = \frac{1}{6}$

6 $k = \frac{3}{4}$ or 0.75

7 $k = \frac{4}{255}$

8 a $k = \frac{1}{4}$ or 0.25

b

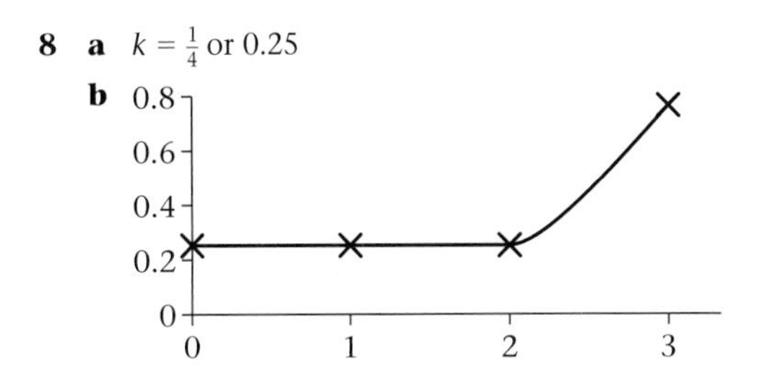

Exercise 3B

1
$$F(x) = \begin{cases} 0 & x < 0 \\ \dfrac{3x^3}{24} & 0 \leqslant x \leqslant 2 \\ 1 & x > 2 \end{cases}$$

2
$$F(x) = \begin{cases} 0 & x < 1 \\ x - \dfrac{x^2}{8} - \dfrac{7}{8} & 1 \leqslant x \leqslant 3 \\ 1 & x > 3 \end{cases}$$

3
$$F(x) = \begin{cases} 0 & x \leqslant 0 \\ \dfrac{x^2}{18} & 0 < x < 3 \\ \dfrac{2x}{3} - \dfrac{x^2}{18} - 1 & 3 \leqslant x \leqslant 6 \\ 1 & x > 6 \end{cases}$$

4 a f(x)

5k, 4k, 3k, 2k, 1k, 0; x: 0 1 2 3 4 5

b $k = \frac{1}{9}$

c
$$F(x) = \begin{cases} 0 & x \leqslant 0 \\ \dfrac{x}{9} & 0 < x < 3 \\ \dfrac{x^2}{9} - \dfrac{5x}{9} + 1 & 3 \leqslant x \leqslant 5 \\ 1 & x > 5 \end{cases}$$

5
$$f(x) = \begin{cases} \dfrac{2x}{5} & 2 \leqslant x \leqslant 3 \\ 0 & \text{otherwise} \end{cases}$$

6 a 0.75 **b** 0.75 **c** 0.5

7 a
$$F(x) = \begin{cases} 0 & x < 0 \\ \dfrac{x^3}{8} & 0 \leqslant x < 2 \\ 1 & x \geqslant 2 \end{cases}$$

b $\frac{1}{8}$

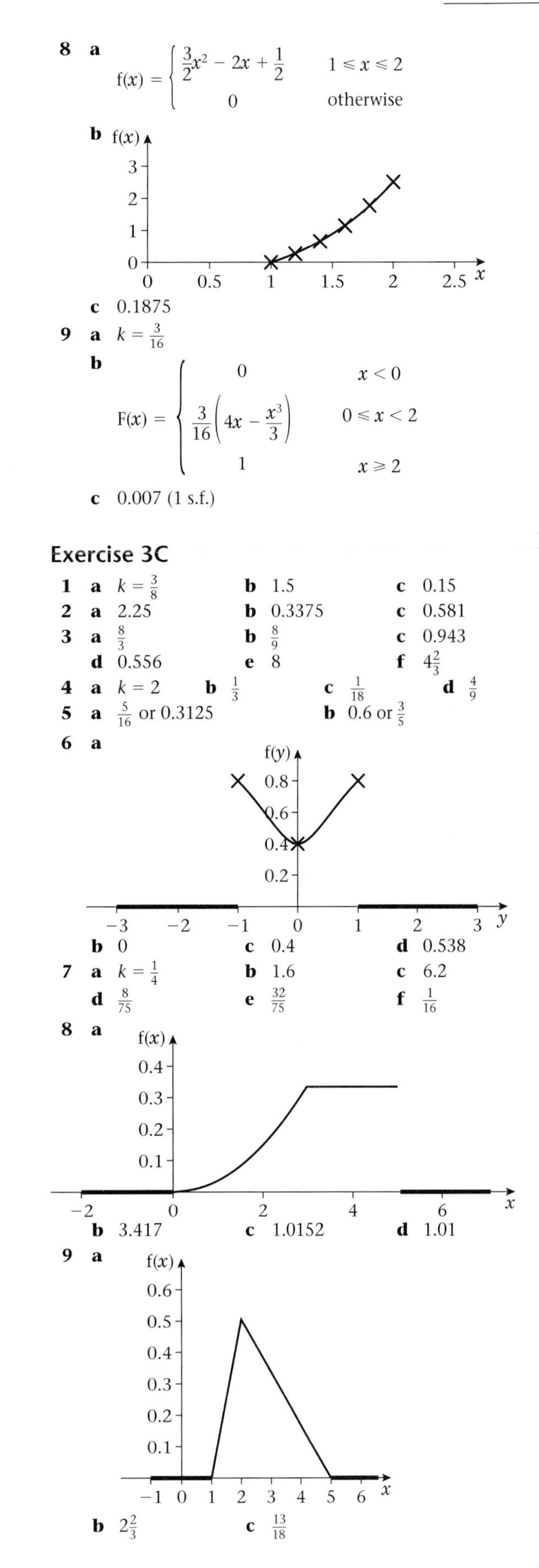

8 a
$$f(x) = \begin{cases} \dfrac{3}{2}x^2 - 2x + \dfrac{1}{2} & 1 \leqslant x \leqslant 2 \\ 0 & \text{otherwise} \end{cases}$$

b f(x)

3, 2, 1, 0; x: 0 0.5 1 1.5 2 2.5

c 0.1875

9 a $k = \frac{3}{16}$

b
$$F(x) = \begin{cases} 0 & x < 0 \\ \dfrac{3}{16}\left(4x - \dfrac{x^3}{3}\right) & 0 \leqslant x < 2 \\ 1 & x \geqslant 2 \end{cases}$$

c 0.007 (1 s.f.)

Exercise 3C

1 a $k = \frac{3}{8}$ **b** 1.5 **c** 0.15

2 a 2.25 **b** 0.3375 **c** 0.581

3 a $\frac{8}{3}$ **b** $\frac{8}{9}$ **c** 0.943 **d** 0.556 **e** 8 **f** $4\frac{2}{3}$

4 a $k = 2$ **b** $\frac{1}{3}$ **c** $\frac{1}{18}$ **d** $\frac{4}{9}$

5 a $\frac{5}{16}$ or 0.3125 **b** 0.6 or $\frac{3}{5}$

6 a f(y)

0.8, 0.6, 0.4, 0.2; y: −3 −2 −1 0 1 2 3

b 0 **c** 0.4 **d** 0.538

7 a $k = \frac{1}{4}$ **b** 1.6 **c** 6.2 **d** $\frac{8}{75}$ **e** $\frac{32}{75}$ **f** $\frac{1}{16}$

8 a f(x)

0.4, 0.3, 0.2, 0.1; x: −2 0 2 4 6

b 3.417 **c** 1.0152 **d** 1.01

9 a f(x)

0.6, 0.5, 0.4, 0.3, 0.2, 0.1; x: −1 0 1 2 3 4 5 6

b $2\frac{2}{3}$ **c** $\frac{13}{18}$

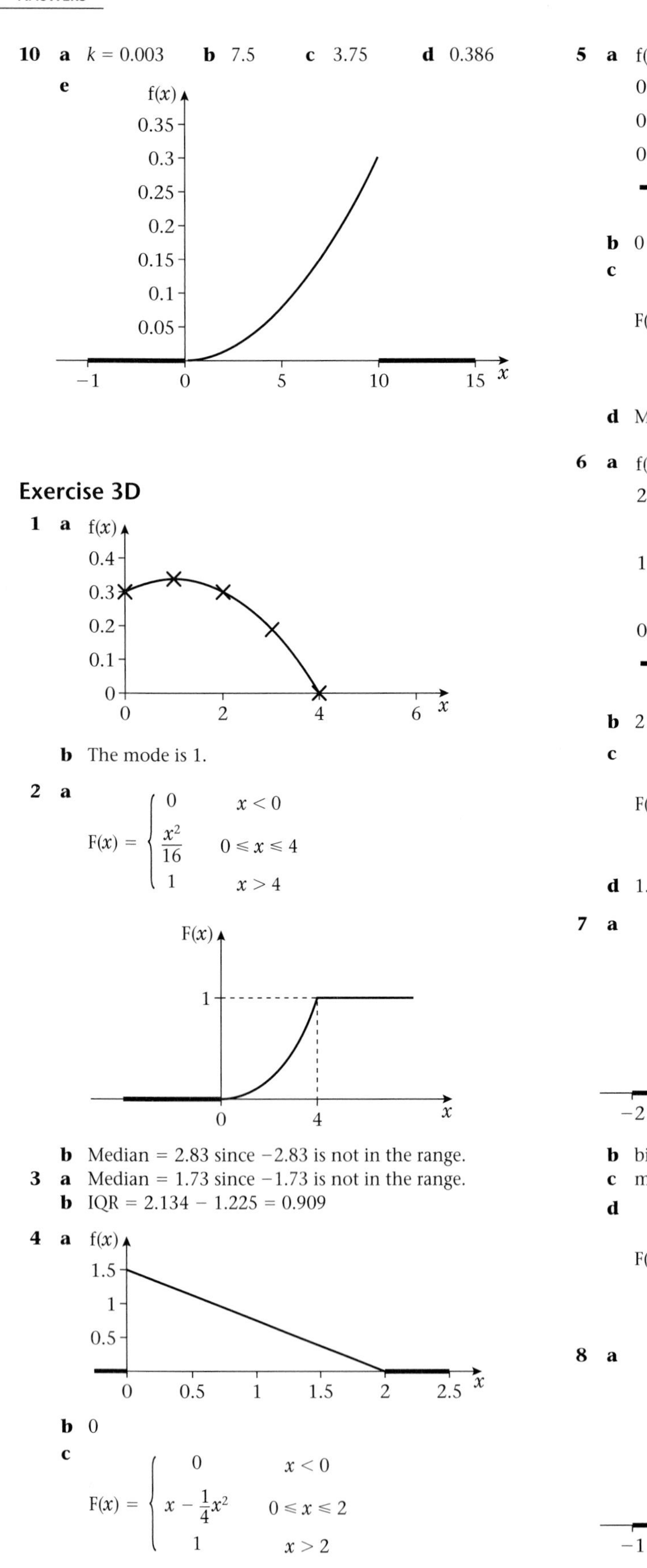

10 **a** $k = 0.003$ **b** 7.5 **c** 3.75 **d** 0.386

e

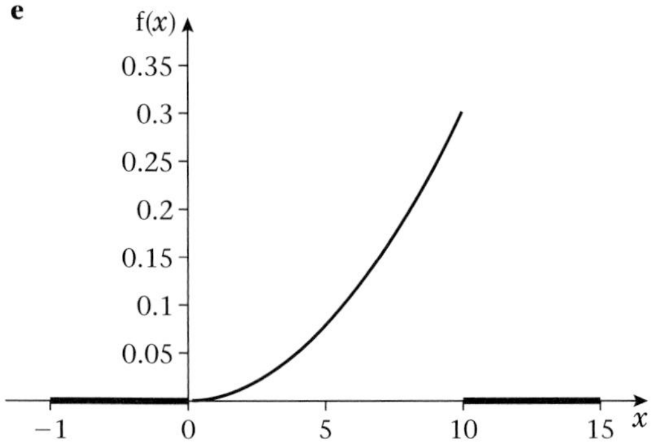

Exercise 3D

1 **a**

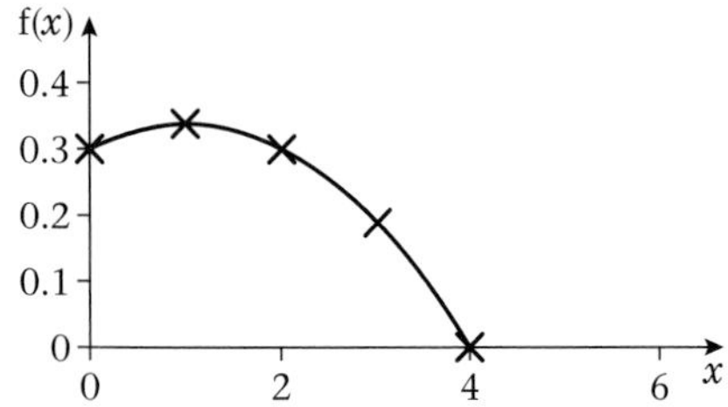

b The mode is 1.

2 **a**

$$F(x) = \begin{cases} 0 & x < 0 \\ \dfrac{x^2}{16} & 0 \leqslant x \leqslant 4 \\ 1 & x > 4 \end{cases}$$

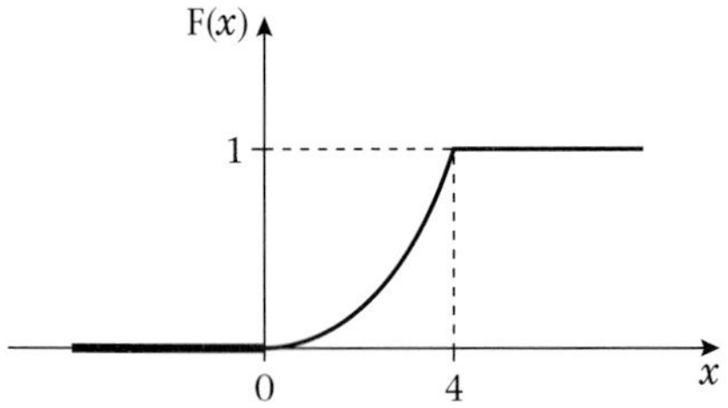

b Median = 2.83 since −2.83 is not in the range.

3 **a** Median = 1.73 since −1.73 is not in the range.

b IQR = 2.134 − 1.225 = 0.909

4 **a**

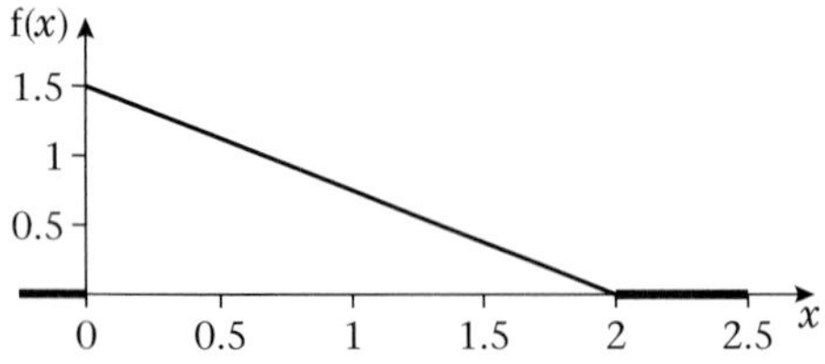

b 0

c

$$F(x) = \begin{cases} 0 & x < 0 \\ x - \frac{1}{4}x^2 & 0 \leqslant x \leqslant 2 \\ 1 & x > 2 \end{cases}$$

d Median = $2 - \sqrt{2}$ as $2 + \sqrt{2}$ is not in range.

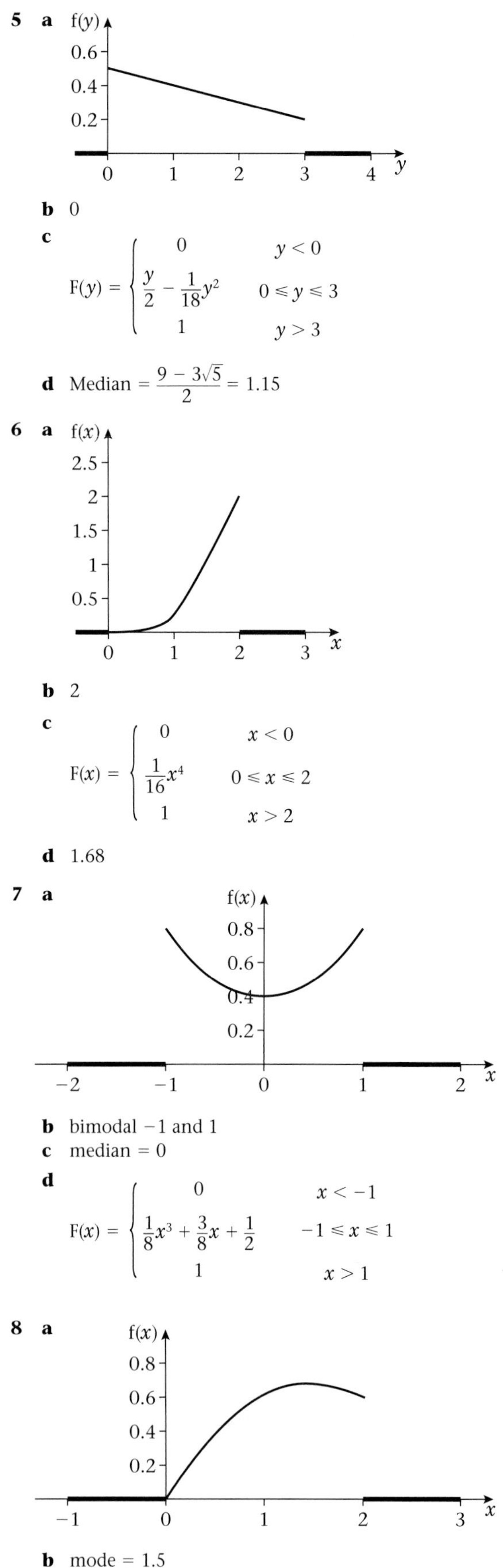

5 **a**

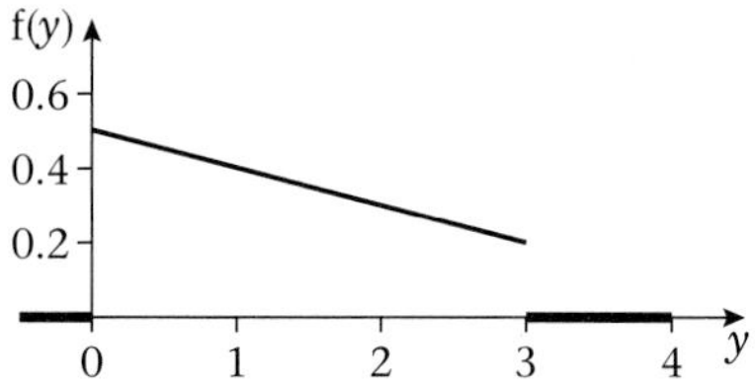

b 0

c

$$F(y) = \begin{cases} 0 & y < 0 \\ \dfrac{y}{2} - \dfrac{1}{18}y^2 & 0 \leqslant y \leqslant 3 \\ 1 & y > 3 \end{cases}$$

d Median $= \dfrac{9 - 3\sqrt{5}}{2} = 1.15$

6 **a**

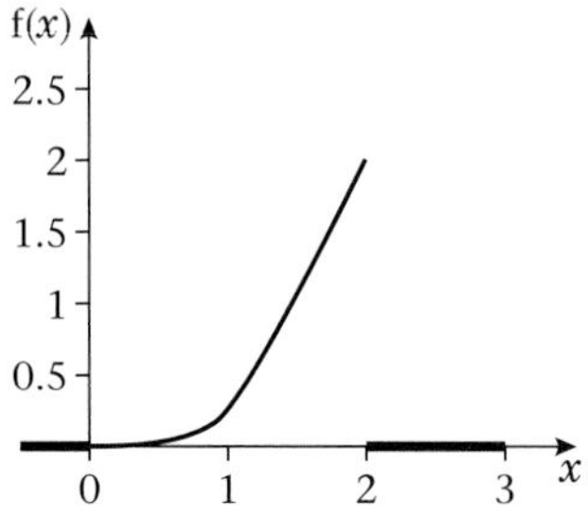

b 2

c

$$F(x) = \begin{cases} 0 & x < 0 \\ \dfrac{1}{16}x^4 & 0 \leqslant x \leqslant 2 \\ 1 & x > 2 \end{cases}$$

d 1.68

7 **a**

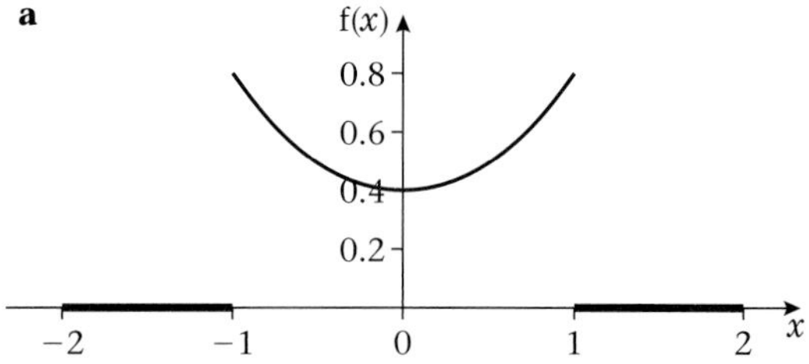

b bimodal −1 and 1

c median = 0

d

$$F(x) = \begin{cases} 0 & x < -1 \\ \frac{1}{8}x^3 + \frac{3}{8}x + \frac{1}{2} & -1 \leqslant x \leqslant 1 \\ 1 & x > 1 \end{cases}$$

8 **a**

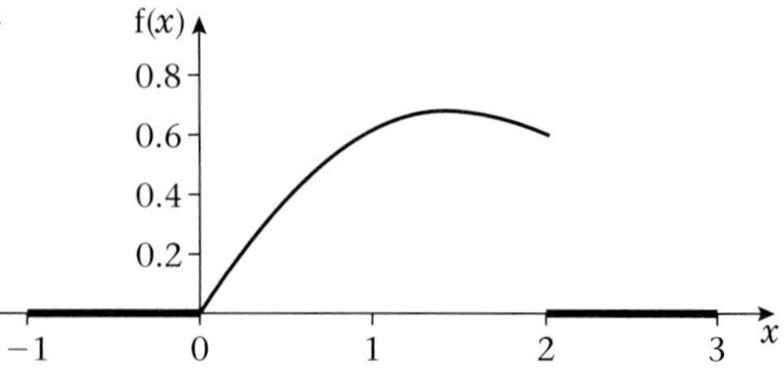

b mode = 1.5

c

$$F(x) = \begin{cases} 0 & x < 0 \\ \frac{9}{20}x^2 - \frac{1}{10}x^3 & 0 \leqslant x \leqslant 2 \\ 1 & x > 2 \end{cases}$$

d Since 0.5 lies between 0.495 and 0.501 the median lies between 1.23 and 1.24.

9 a

$$f(x) = \begin{cases} \frac{1}{4}x & 1 \leqslant x \leqslant 3 \\ 0 & \text{otherwise} \end{cases}$$

b mode = 3

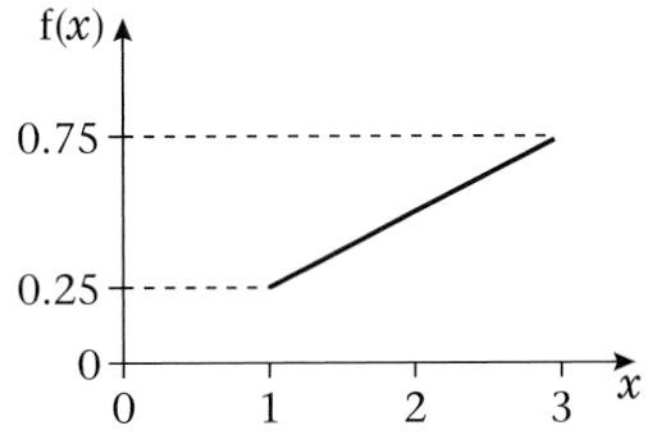

c $\sqrt{5}$

d lower quartile = $\sqrt{3}$
upper quartile = $\sqrt{7}$

10 a

$$f(x) = \begin{cases} 12x^2(1 - x) & 0 \leqslant x \leqslant 1 \\ 0 & \text{otherwise} \end{cases}$$

b mode = $\frac{2}{3}$

c 0.2853

11 a

$$F(x) = \begin{cases} 0 & x < 0 \\ \frac{w^4}{5^5}(25 - 4w) & 0 \leqslant x \leqslant 5 \\ 1 & x > 5 \end{cases}$$

b 0.650

12 a

$$F(x) = \begin{cases} 0 & x < 0 \\ \frac{x}{4} & 0 \leqslant x < 1 \\ \frac{x^4}{20} + \frac{1}{5} & 1 \leqslant x \leqslant 2 \\ 1 & x > 2 \end{cases}$$

b median = 1.57 IQR = 0.821

Mixed exercise 3E

1 a $5\frac{1}{3}$ **b** 2.89 **c** $\frac{5}{12}$

d $\frac{128}{243}$ **e** 0.5

2 a $\frac{1}{3}$ **b** $\frac{1}{18}$ **c** $\frac{4}{18} = \frac{2}{9}$

d

$$F(x) = \begin{cases} 0 & x < 0 \\ 2x - x^2 & 0 \leqslant x \leqslant 1 \\ 1 & x > 1 \end{cases}$$

e median = 0.293 as 1.71 is not in the range

3 a $k = \frac{1}{2}$ **b** 0.375

c median = 1.62 as −0.618 is not in the range

d

$$f(y) = \begin{cases} y - \frac{1}{2} & 1 \leqslant y \leqslant 2 \\ 0 & \text{otherwise} \end{cases}$$

4 a 0.648

b median = 2.55 as −2.55 is not in the range

c

$$f(x) = \begin{cases} \frac{2x}{5} & 2 \leqslant x \leqslant 3 \\ 0 & \text{otherwise} \end{cases}$$

d $\frac{38}{15}$ **e** mode = 3

5 a $k = \frac{3}{8}$ **b** 1.5

c

$$F(x) = \begin{cases} 0 & x < 0 \\ \frac{x^3}{8} & 0 \leqslant x \leqslant 2 \\ 1 & x > 2 \end{cases}$$

d 1.59 **e** mode = 2

6 a $k = \frac{3}{62}$

b

$$F(y) = \begin{cases} 0 & y < 1 \\ \frac{y^3}{62} + \frac{3y^2}{62} + \frac{3y}{31} - \frac{5}{31} & 1 \leqslant y \leqslant 3 \\ 1 & y > 3 \end{cases}$$

c $\frac{11}{31}$

7 a

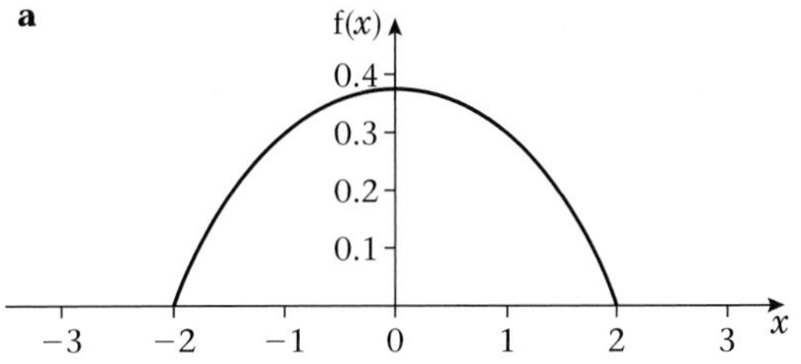

b mode = 0

c

$$F(x) = \begin{cases} 0 & x < -2 \\ \frac{12x}{32} - \frac{x^3}{32} + \frac{1}{2} & -2 \leqslant x \leqslant 2 \\ 1 & x > 2 \end{cases}$$

d $\frac{35}{128}$ or 0.273

8 a $\frac{26}{21}$

b

$$F(x) = \begin{cases} 0 & x < 0 \\ \frac{x}{3} & 0 \leqslant x < 1 \\ \frac{2x^3}{21} + \frac{5}{21} & 1 \leqslant x \leqslant 2 \\ 1 & x > 2 \end{cases}$$

c median = 1.40

9 a $k = \frac{1}{2}$ **b** $\frac{7}{3}$

c

$$F(x) = \begin{cases} 0 & x < 1 \\ \frac{x^2}{4} - \frac{x}{2} + \frac{1}{4} & 1 \leqslant x \leqslant 3 \\ 1 & x > 3 \end{cases}$$

d Since 0.5 lies between 0.49 and 0.5625 the median is between 2.4 and 2.5

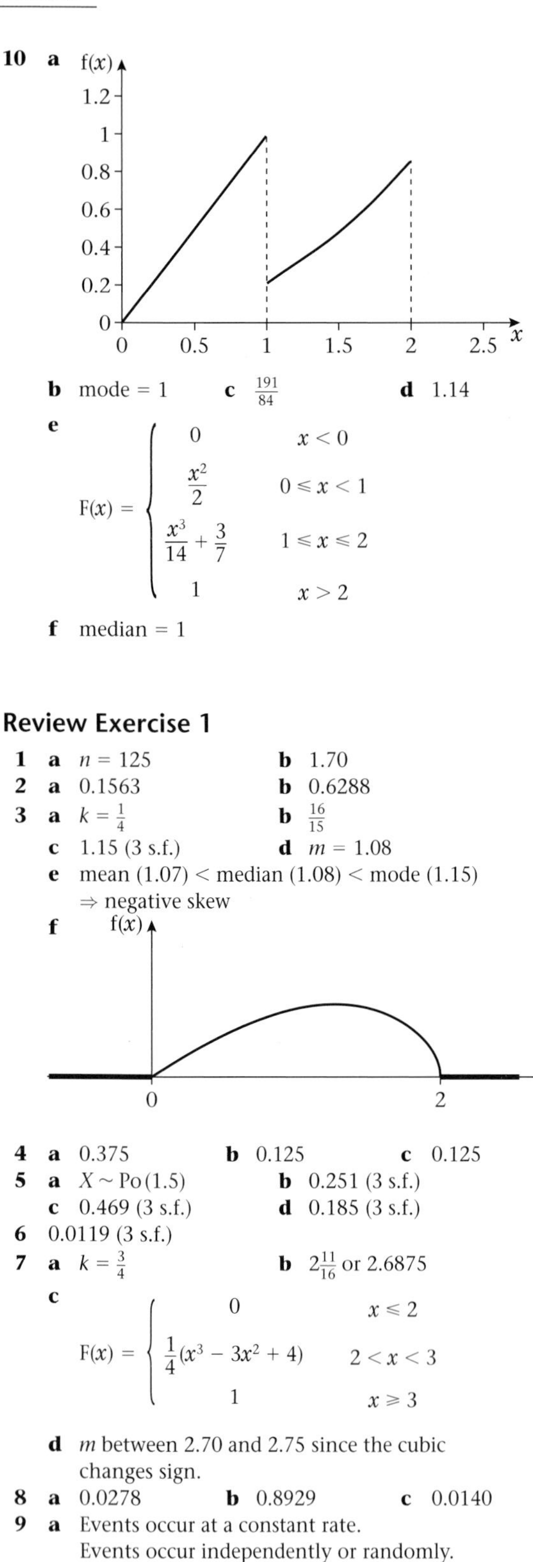

10 **a** (graph of f(x))

b mode = 1 **c** $\frac{191}{84}$ **d** 1.14

e

$$F(x) = \begin{cases} 0 & x < 0 \\ \dfrac{x^2}{2} & 0 \leqslant x < 1 \\ \dfrac{x^3}{14} + \dfrac{3}{7} & 1 \leqslant x \leqslant 2 \\ 1 & x > 2 \end{cases}$$

f median = 1

Review Exercise 1

1 **a** $n = 125$ **b** 1.70

2 **a** 0.1563 **b** 0.6288

3 **a** $k = \frac{1}{4}$ **b** $\frac{16}{15}$

c 1.15 (3 s.f.) **d** $m = 1.08$

e mean (1.07) < median (1.08) < mode (1.15) ⇒ negative skew

f (graph of f(x))

4 **a** 0.375 **b** 0.125 **c** 0.125

5 **a** $X \sim \text{Po}(1.5)$ **b** 0.251 (3 s.f.)

c 0.469 (3 s.f.) **d** 0.185 (3 s.f.)

6 0.0119 (3 s.f.)

7 **a** $k = \frac{3}{4}$ **b** $2\frac{11}{16}$ or 2.6875

c

$$F(x) = \begin{cases} 0 & x \leqslant 2 \\ \frac{1}{4}(x^3 - 3x^2 + 4) & 2 < x < 3 \\ 1 & x \geqslant 3 \end{cases}$$

d m between 2.70 and 2.75 since the cubic changes sign.

8 **a** 0.0278 **b** 0.8929 **c** 0.0140

9 **a** Events occur at a constant rate.
Events occur independently or randomly.
Events occur singly.

b **i** 0.134 (3 s.f.) **ii** 0.715 (3 s.f.)

c 0.149

10 **a** $k = \frac{1}{18}$ **b** 0.705 $\left[\text{or } \frac{203}{288}\right]$

c

$$f(y) = \begin{cases} 0 & \text{otherwise} \\ \frac{1}{9}(2y^3 + y) & 1 \leqslant y \leqslant 2 \end{cases}$$

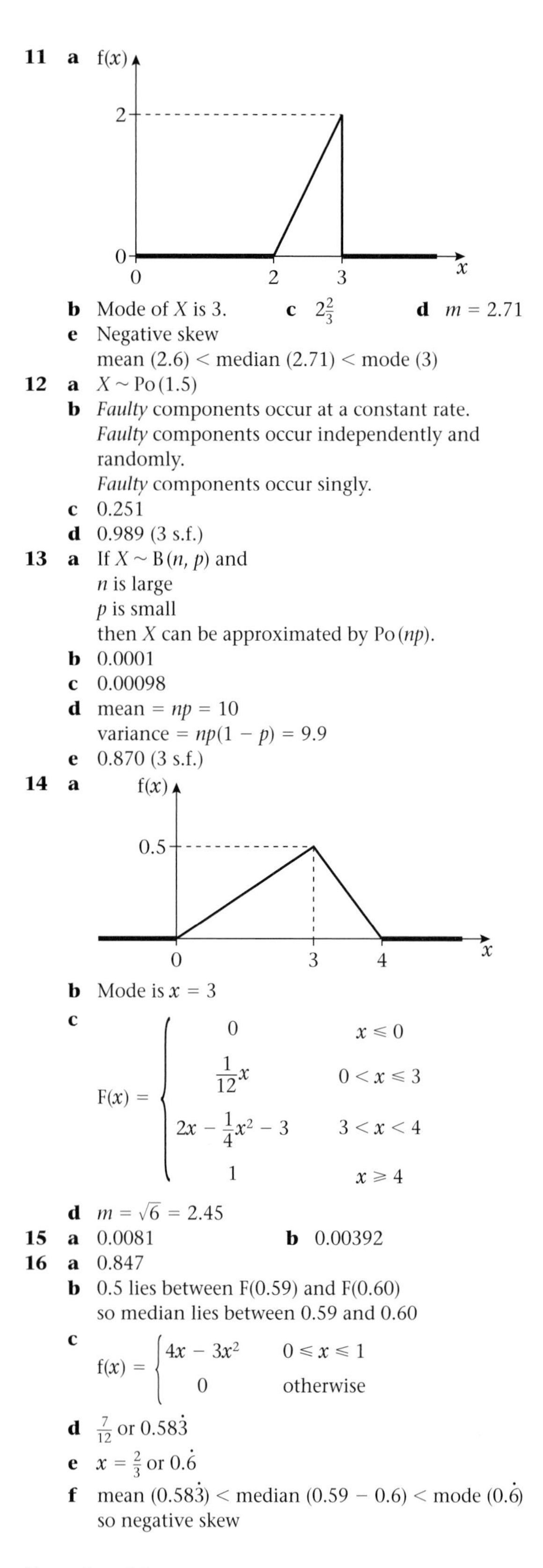

11 **a** (graph of f(x))

b Mode of X is 3. **c** $2\frac{2}{3}$ **d** $m = 2.71$

e Negative skew
mean (2.6) < median (2.71) < mode (3)

12 **a** $X \sim \text{Po}(1.5)$

b *Faulty* components occur at a constant rate.
Faulty components occur independently and randomly.
Faulty components occur singly.

c 0.251

d 0.989 (3 s.f.)

13 **a** If $X \sim \text{B}(n, p)$ and
n is large
p is small
then X can be approximated by $\text{Po}(np)$.

b 0.0001

c 0.00098

d mean $= np = 10$
variance $= np(1 - p) = 9.9$

e 0.870 (3 s.f.)

14 **a** (graph of f(x))

b Mode is $x = 3$

c

$$F(x) = \begin{cases} 0 & x \leqslant 0 \\ \dfrac{1}{12}x & 0 < x \leqslant 3 \\ 2x - \dfrac{1}{4}x^2 - 3 & 3 < x < 4 \\ 1 & x \geqslant 4 \end{cases}$$

d $m = \sqrt{6} = 2.45$

15 **a** 0.0081 **b** 0.00392

16 **a** 0.847

b 0.5 lies between F(0.59) and F(0.60) so median lies between 0.59 and 0.60

c

$$f(x) = \begin{cases} 4x - 3x^2 & 0 \leqslant x \leqslant 1 \\ 0 & \text{otherwise} \end{cases}$$

d $\frac{7}{12}$ or $0.58\dot{3}$

e $x = \frac{2}{3}$ or $0.\dot{6}$

f mean $(0.58\dot{3})$ < median (0.59 – 0.6) < mode $(0.\dot{6})$ so negative skew

Exercise 4A

1 **a** 0.4 **b** 0.6

2 **a** $k = 12.6$ **b** 0.39

3 **a** $k = \frac{1}{8}$ **b** 0.6875

4

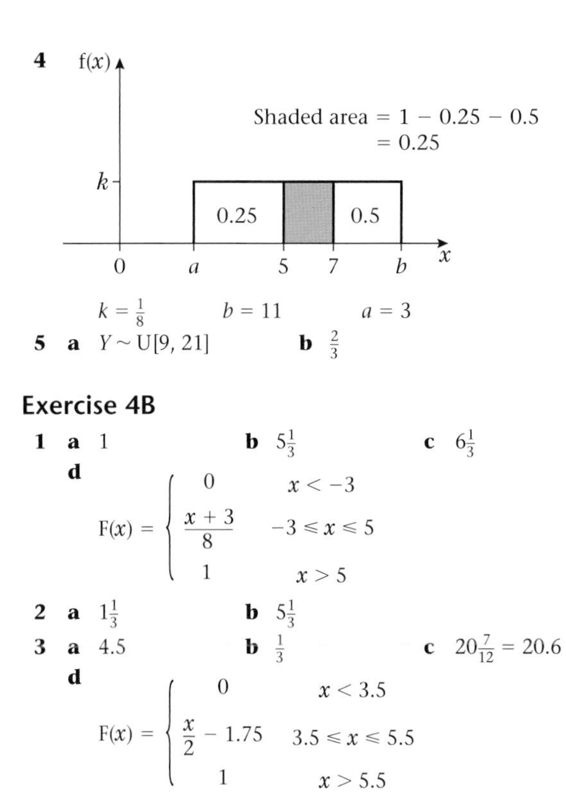

$k = \frac{1}{8}$ $\quad b = 11 \quad a = 3$

5 **a** $Y \sim \text{U}[9, 21]$ **b** $\frac{2}{3}$

Exercise 4B

1 **a** 1 **b** $5\frac{1}{3}$ **c** $6\frac{1}{3}$

d
$$F(x) = \begin{cases} 0 & x < -3 \\ \dfrac{x+3}{8} & -3 \leqslant x \leqslant 5 \\ 1 & x > 5 \end{cases}$$

2 **a** $1\frac{1}{3}$ **b** $5\frac{1}{3}$

3 **a** 4.5 **b** $\frac{1}{3}$ **c** $20\frac{7}{12} = 20.6$

d
$$F(x) = \begin{cases} 0 & x < 3.5 \\ \dfrac{x}{2} - 1.75 & 3.5 \leqslant x \leqslant 5.5 \\ 1 & x > 5.5 \end{cases}$$

4 $a = -1$ and $b = 3$

5 $E(X) = \dfrac{5 + (-1)}{2}$
$= 2$
$\text{Var}(X) = \dfrac{(5 - (-1))^2}{12}$
$= 3$
$E(Y) = 4E(X) - 6$
$= 8 - 6$
$= 2$
$\text{Var}(Y) = 16\,\text{Var}(X)$
$= 48$

6 $E(X^2) = 25\frac{1}{12}$

7 **a** 0.2 **b** 0.5 **c** $\frac{1}{12}$

Mixed exercise 4C

1 **a**

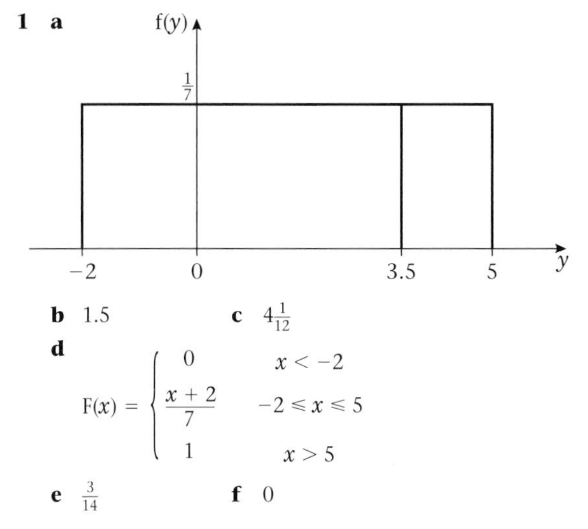

b 1.5 **c** $4\frac{1}{12}$

d
$$F(x) = \begin{cases} 0 & x < -2 \\ \dfrac{x+2}{7} & -2 \leqslant x \leqslant 5 \\ 1 & x > 5 \end{cases}$$

e $\frac{3}{14}$ **f** 0

2 **a** $k = 1$ **b** 0.2 **c** -1.5 **d** $2\frac{1}{12}$

e
$$F(x) = \begin{cases} 0 & x < -4 \\ \dfrac{x+4}{5} & -4 \leqslant x \leqslant 1 \\ 1 & x > 1 \end{cases}$$

3 **a** $b = 5 \quad a = -1$

b 0.533

4 **a**

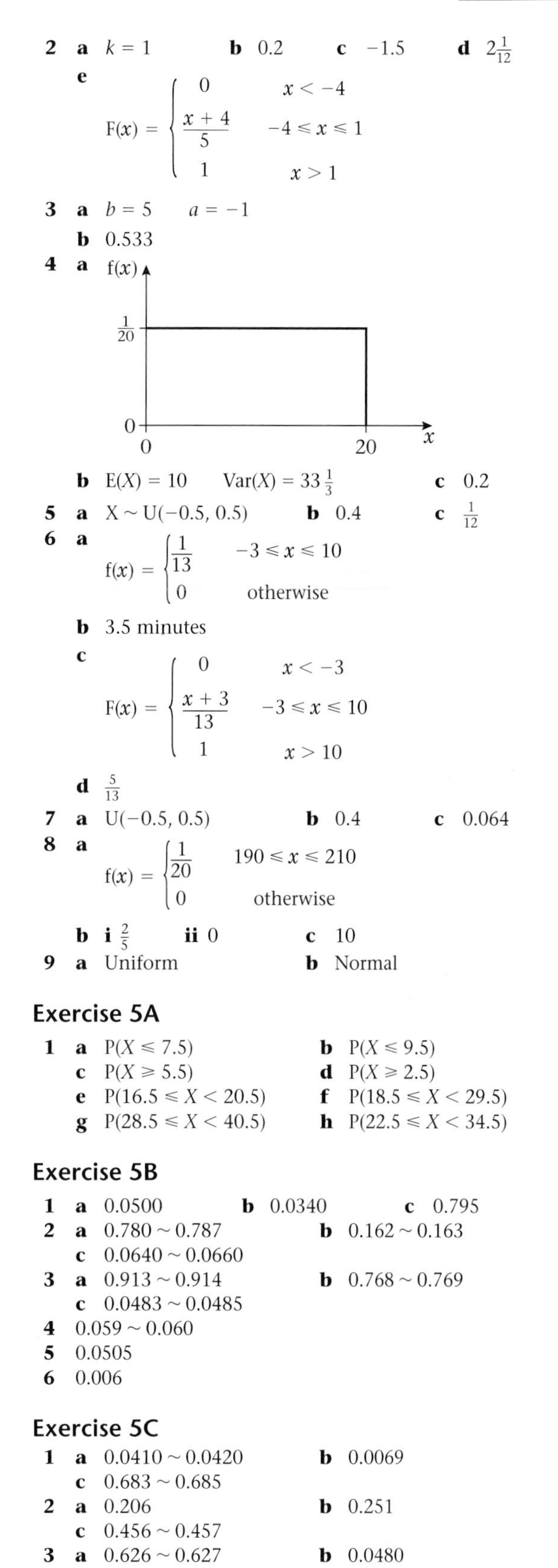

b $E(X) = 10 \quad \text{Var}(X) = 33\frac{1}{3}$ **c** 0.2

5 **a** $X \sim \text{U}(-0.5, 0.5)$ **b** 0.4 **c** $\frac{1}{12}$

6 **a**
$$f(x) = \begin{cases} \dfrac{1}{13} & -3 \leqslant x \leqslant 10 \\ 0 & \text{otherwise} \end{cases}$$

b 3.5 minutes

c
$$F(x) = \begin{cases} 0 & x < -3 \\ \dfrac{x+3}{13} & -3 \leqslant x \leqslant 10 \\ 1 & x > 10 \end{cases}$$

d $\frac{5}{13}$

7 **a** U(−0.5, 0.5) **b** 0.4 **c** 0.064

8 **a**
$$f(x) = \begin{cases} \dfrac{1}{20} & 190 \leqslant x \leqslant 210 \\ 0 & \text{otherwise} \end{cases}$$

b **i** $\frac{2}{5}$ **ii** 0 **c** 10

9 **a** Uniform **b** Normal

Exercise 5A

1 **a** $P(X \leqslant 7.5)$ **b** $P(X \leqslant 9.5)$
c $P(X \geqslant 5.5)$ **d** $P(X \geqslant 2.5)$
e $P(16.5 \leqslant X < 20.5)$ **f** $P(18.5 \leqslant X < 29.5)$
g $P(28.5 \leqslant X < 40.5)$ **h** $P(22.5 \leqslant X < 34.5)$

Exercise 5B

1 **a** 0.0500 **b** 0.0340 **c** 0.795

2 **a** 0.780 ~ 0.787 **b** 0.162 ~ 0.163
c 0.0640 ~ 0.0660

3 **a** 0.913 ~ 0.914 **b** 0.768 ~ 0.769
c 0.0483 ~ 0.0485

4 0.059 ~ 0.060

5 0.0505

6 0.006

Exercise 5C

1 **a** 0.0410 ~ 0.0420 **b** 0.0069
c 0.683 ~ 0.685

2 **a** 0.206 **b** 0.251
c 0.456 ~ 0.457

3 **a** 0.626 ~ 0.627 **b** 0.0480
c 0.480 ~ 0.481

4 **a** 0.0590 ~ 0.0600 **b** 0.825 ~ 0.826
c 0.311 ~ 0.312

5 **a** 1 day **b** 4 or 5 days **c** 8 days

Mixed exercise 5D

1 a 0.8159 b 0.135 ~ 0.136
2 0.0778
3 a 0.0262 b 0.2149
4 a 0.1849 b 0.8576 c 0.946
5 a 0.104 b 0.8153

Exercise 6A

1 a A census is when **every member of a population** is used.
  b ANY TWO FROM:
    It is unbiased.
    It gives an accurate, reliable answer.
    It looks at every single member of the population.
  c ANY TWO FROM:
    It can take a long time to do.
    It is often costly.
    It is not easy to ensure that every member of the population is taken into account.
2 **a** is an infinite population.
  b and **c** are finite populations.
3 a EITHER: A sample is a subset of the population.
    OR: A sample consists of a selected group of the members of the population.
  b ANY ONE FROM:
    It may be biased.
    It may be subject to natural variation.
  c ANY TWO FROM:
    It is generally cheaper.
    Data is often easier to get.
    It generally takes less time.
    It avoids testing to destruction.
4 a ANY ONE FROM:
    It would be expensive.
    It would be time consuming.
    It would be difficult.
  b A list of residents.
  c A resident.
5 The climbing ropes would all be destroyed.
6 a ANY ONE FROM:
    It will be easier.
    It will be quicker.
    It will be cheaper.
  b Customer.
  c Advantages:
    ANY ONE FROM:
    It will be quick to do.
    It will be easy to do.
    It will not cost too much.
    PLUS
    Disadvantages:
    ANY ONE FROM:
    Not everyone's views will be known.
    It might be biased.
7 a All the mechanics in the garage.
  b Everyone's views will be known.
8 a If a census were used all the computers would be destroyed.
  b The list of unique serial numbers.
  c A computer.

Exercise 6B

1 This mean is from the values of a sample so it is a statistic.
2 **i** and **ii** are statistics.
  iii is not a statistic since it uses μ.
3 a All the hairdressers who work for the chain of hairdressing shops.
    The proportion of p of the staff happy to wear overalls.
  b This is a binomial distribution since we are only interested in two options – whether or not the hairdressers are happy to wear the overalls.
4 a Po(3) b 0.1991
5 a $\sigma = 318.75$ $\mu = 32.5$
  b (50, 50)
    (50, 20) (20, 50)
    (50, 10) (10, 50)
    (20, 20)
    (20, 10) (10, 20)
    (10, 10)
  c

$\bar{X}$	50	35	30	20	15	10
$P(\bar{X})$	0.25	0.25	0.25	0.0625	0.125	0.0625

6 a Mean = 19.4 Variance = 16.04
  b (16, 16)
    (16, 20) (20, 16)
    (16, 30) (30, 16)
    (30, 30)
    (30, 20) (20, 30)
    (20, 20)
  c

$\bar{X}$	16	18	20	23	25	30
$P(\bar{X})$	0.16	0.4	0.25	0.08	0.1	0.01

7 a Mean = 2.6 Variance = 0.24
  b (3, 3, 3)
    (3, 3, 2) (3, 2, 3) (2, 3, 3)
    (3, 2, 2) (2, 3, 2) (2, 2, 3)
    (2, 2, 2)
  c

$\bar{X}$	3	$2\frac{2}{3}$	$2\frac{1}{3}$	2
$P(\bar{X})$	0.216	0.432	0.288	0.064

  d

M	3	2
$P(M)$	0.648	0.352

  e

N	3	2
$P(N)$	0.648	0.352

Mixed exercise 6C

1 a A list of all the patients on the surgery books.
  b A patient.
2 a ANY TWO FROM:
    It would take too long.
    It could cost too much.
    It could be difficult to get hold of all members.
  b A list of all members of the gym.
  c A member of the gym.
3 a A sampling frame has to be some sort of list – it may not be possible to list a population.
  b A sample is usually easier to do, quicker to do and not as costly as a census.

4 a A statistic is a quantity calculated solely from the observations of a sample.

b i is a statistic **ii** is not a statistic as it depends on the value μ.

5 a The light bulbs would all be destroyed.

b A light bulb.

6 a ANY TWO FROM:
It is quicker to do.
It is cheaper to do.
It is easier to do.

b It can be biased. OR it is subject to natural variations.

c A numbered list of all 400 call-centre operatives.

d A call-centre operative.

e Yes, because he is using only the values from a sample. There are no parameters.

7 ANY TWO FROM:
It takes into account everyone's views.
It is unbiased.
To take a sample when the population is only 10 would be silly.

8 **a** and **b** are statistics
c and **d** are not statistics since they involve a population parameter.

9 a Mean = $9\frac{1}{6}$ Variance = 28.47

b (5, 5) (10, 10) (20, 20)
(5, 10) (10, 5) (5, 20) (20, 5)
(10, 20) (20, 10)

c

$\bar{Y}$	5	7.5	10	12.5	15	20
$P(\bar{Y})$	$\frac{1}{4}$	$\frac{1}{3}$	$\frac{1}{9}$	$\frac{1}{6}$	$\frac{1}{9}$	$\frac{1}{36}$

10 a (6, 6, 6)
(6, 6, 10) (6, 10, 6) (10, 6, 6)
(6, 10, 10) (10, 6, 10) (10, 10, 6)
(10, 10, 10)

b

N	6	10
$P(N)$	0.648	0.352

c

M	3	2
$P(M)$	0.648	0.352

Exercise 7A

1 a This is an assumption made about a population parameter that we test using evidence from a sample.

b The null hypothesis is what we assume to be correct and the alternative hypothesis is what we conclude if our assumption is wrong.

c Null hypothesis = H_0 Alternative hypothesis = H_1

2 a The test statistic is N – the number of sixes.

b H_0: $p = \frac{1}{6}$

c H_1: $p > \frac{1}{6}$

3 a The test statistic is N – the number of times you get a head.

b H_0: $p = \frac{1}{2}$

c H_1: $p \neq \frac{1}{2}$

4 a The test statistic is the number of accidents (in a given month or other specified time period).

b H_0: $\lambda = 4$

c Change H_1: $\lambda \neq 4$ (2 tail); or Decrease H_1: $\lambda < 4$ or Increase H_1: $\lambda > 4$ (both one tail).

5 a A two tail test would be best. The support could get better or could get worse.

b H_0: $p = 0.4$
H_1: $p \neq 0.4$

6 a A suitable test statistic is p – the proportion of faulty articles in a batch.

b H_0: $p = 0.1$
H_1: $p < 0.1$

c If the probability of the proportion being 0.09 or less is 5% or less the null hypothesis is rejected.

7 a The test statistic is M – the number of times Hajdra gets a 1.

b If N is 4 or 5 then the null hypothesis would be rejected, since P(4 or more) = 1.6% < 5%.

Exercise 7B

1 $0.0781 > 0.05$
There is insufficient evidence to reject H_0.

2 $0.0464 < 0.05$
There is sufficient evidence to reject H_0 so $p < 0.04$.

3 $0.0480 < 0.05$
There is sufficient evidence to reject H_0 so $p > 0.30$.

4 $0.0049 < 0.01$
There is sufficient evidence to reject H_0 so $p < 0.45$.

5 $0.0494 > 0.025$ (two-tailed)
There is insufficient evidence to reject H_0 so there is no reason to doubt $p = 0.5$.

6 $0.0526 > 0.05$
There is insufficient evidence to reject H_0 so there is no reason to doubt $p = 0.28$.

7 $0.0020 < 0.05$
There is sufficient evidence to reject H_0 so $p > 0.32$.

8 $0.0424 < 0.05$
There is sufficient evidence to reject H_0 so $\lambda < 8$.

9 $0.0430 > 0.01$ (1% sig. level)
There is insufficient evidence to reject H_0.

10 $0.1905 > 0.05$
There is insufficient evidence to reject H_0.

11 $0.1875 > 0.05$
There is insufficient evidence to reject H_0 (not significant).
There is insufficient evidence to suggest that people prefer Supergold to butter.

12 $0.0577 > 0.025$ (two-tailed)
There is insufficient evidence to reject H_0.

13 $0.3813 > 0.05$
There is insufficient evidence to reject H_0 (not significant).
There is no evidence that the probability is less than $\frac{1}{6}$.
There is no evidence that the die is biased.

14 a Distribution B(n, 0.68)
Fixed number of trials.
Outcomes of trials are independent.
There are two outcomes success and failure.
The probability of success is constant.

b $0.0155 < 0.05$
There is insufficient evidence to reject H_0 so $p < 0.68$.
The treatment is not as effective as is claimed.

15 $0.1321 > 0.05$
There is insufficient evidence to reject H_0 (not significant).
There is no evidence of a decrease in the number of lorries passing the gates in a lunch hour.

16 $0.1242 > 0.05$
There is insufficient evidence to reject H_0 (not significant).
There is no evidence that the new schedules have increased the number of times the bus is late.

Exercise 7C

1 The critical value is $x = 5$ and the critical region is $X \geqslant 5$ since $P(X \geqslant 5) = 0.0328 < 0.05$.
2 The critical value is $x = 0$ and the critical region is $X = 0$.
3 The critical region is $X \geqslant 13$ and $X \leqslant 3$.
4 The critical value is $x = 0$. The critical region is $X = 0$.
5 The critical value is $x = 10$ and the critical region is $X = 10$.
6 Critical region is $X \geqslant 7$.
7 Critical region is $X \geqslant 9$.
8 Critical region is $X \leqslant 2$.
9 Critical region is $X = 0$.
10 Critical region is $X \leqslant 6$.
11 **a** Critical region $X = 0$ and $X \geqslant 8$.
b $X = 8$ is in the critical region. There is enough evidence to reject H_0. The hospital's proportion of complications differs from the national figure.
c 0.0436
12 **a** $H_0: \lambda = 4 \quad H_1: \lambda > 4$
b Critical region is $X \geqslant 9$.
c 8 is not in the critical region. The scientist concluded there was not enough evidence to suggest an increase in the number of hurricanes.
13 There is evidence that the rate of weekly sales has decreased.
14 Reject H_0
There is evidence that the percentage of workers who are absent for at least 1 day per month is less than 20%.

Mixed exercise 7D

1 $0.3222 > 0.05$
There is insufficient evidence to reject H_0.
There is no evidence that the trains are late more often.
2 $0.1321 > 0.025$
There is insufficient evidence to reject H_0.
There is no evidence that the rate at which lorries pass the hospital has changed.
3 $0.1875 > 0.05$
There is insufficient evidence to reject H_0.
There is insufficient evidence that the manufacturer's claim is true.
4 $0.2084 > 0.10$
There is insufficient evidence to reject H_0.
There is no evidence that the rate of tremors has increased.
5 **a** Fixed number; independent trials; two outcomes (pass or fail); p constant for each car.
b 0.16807
c $0.3828 > 0.05$
There is insufficient evidence to reject H_0.
There is no evidence that the garage fails fewer than the national average.
6 **a** A hypothesis test about a population parameter θ tests a null hypothesis H_0, which specifies a particular value for θ, against an alternative hypothesis H_1 which is that θ has increased, decreased or changed. H_1 will indicate if the test is one- or two-tailed.
b You can count the number of cups of tea that were served in a given time interval (30 minutes). (You cannot count how many were not served.)
c Critical region is $X \geqslant 6$.
d 0.0166
7 **a** Critical region is $X \geqslant 13$.
b There is insufficient evidence to reject H_0.
There is no evidence that practice has improved Jacinth's throwing.
8 **a** Critical region $X \leqslant 1$ and $X \geqslant 10$.
b 0.0583
c Accept H_0. There is no evidence that the proportion of defective articles has increased.
9 There is insufficient evidence to reject H_0.
There is no evidence that the claim is wrong.
10 **a** 0.1339
b $0.0426 < 0.05$
There is sufficient evidence to reject H_0.
There is evidence that the rate of hiring caravans has increased.
11 **a** Critical region is $X \leqslant 4$ and $X \geqslant 16$
b 0.0493
c There is insufficient evidence to reject H_0.
There is no evidence to suggest that the proportion of people buying that certain make of computer differs from 0.2.
12 $0.0282 < 0.05$ Reject H_0.
There is evidence that the filter bed is failing to work properly.
13 $0.0011 < 0.05$ Reject H_0.
There is evidence that the rate of sales of onion marmalade has increased after the program.
14 $0.0415 < 0.05$ Reject H_0.
There is evidence that the process is getting worse.
15 Reject H_0.
The new variety is better.

Review Exercise 2

1 **a**

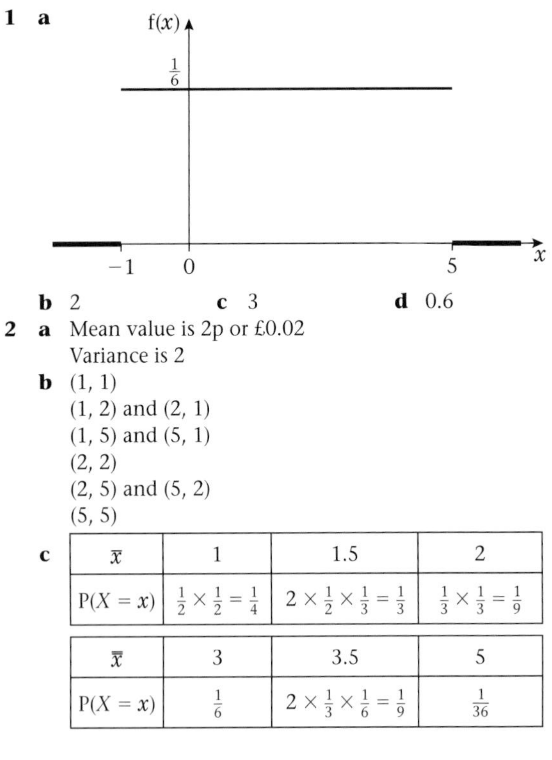

b 2 **c** 3 **d** 0.6
2 **a** Mean value is 2p or £0.02
Variance is 2
b (1, 1)
(1, 2) and (2, 1)
(1, 5) and (5, 1)
(2, 2)
(2, 5) and (5, 2)
(5, 5)
c

$\bar{x}$	1	1.5	2
$P(X = x)$	$\frac{1}{2} \times \frac{1}{2} = \frac{1}{4}$	$2 \times \frac{1}{2} \times \frac{1}{3} = \frac{1}{3}$	$\frac{1}{3} \times \frac{1}{3} = \frac{1}{9}$

$\bar{x}$	3	3.5	5
$P(X = x)$	$\frac{1}{6}$	$2 \times \frac{1}{3} \times \frac{1}{6} = \frac{1}{9}$	$\frac{1}{36}$

3 **a** **i** Evidence that the percentage of pupils that read Deano is not 20%, it is more than 20%.
ii All possible values are 0 or [9, 20] or 0 and 9 or more.
b Combined numbers of Deano readers suggests there is no reason to doubt 20% of pupils read Deano.
c In part **a** we *rejected* H_0.
In part **b** we had *insufficient evidence* to reject H_0.
The results are *different*.
Either sample size matters and *larger samples give more reliable results* or not all pupils are drawn from the same population.

4 **a** $$f(x) = \begin{cases} \frac{1}{4} & 2 \leqslant x \leqslant 6 \\ 0 & \text{otherwise} \end{cases}$$
b $E(x) = 4$
c $\frac{4}{3}$
d $$F(x) = \begin{cases} 0 & x < 2 \\ \frac{1}{4}(x-2) & 2 \leqslant x \leqslant 6 \\ 1 & x > 6 \end{cases}$$
e 0.275

5 **a** Misprints occur randomly and independently.
Misprints occur singly in space.
Misprints occur at a constant rate.
b 0.0821 **c** 0.133 (3 s.f.) **d** 0.07

6 **a** *Element* of the *population.*
b A *list of all* the sampling units.
c *All possible samples* are chosen from a population, the *values* of a statistic and the associated *probabilities* is a sampling distribution.

7 **a** $X \sim B(10, 0.75)$
where X is the random variable 'number of patients who recover when treated'.
b 0.146
c Insufficient evidence to reject H_0.
d 9

8 **a** A census is when *every member* of a *population* is investigated.
b This is destructive testing, so there would be no cookers left to sell if a census were taken.
c A *list* of the serial numbers of the cookers.
d A cooker.

9 Reject H_0.
Dhiriti's claim is supported by sample.

10 **a** **i** 0.644 (3 s.f.) **ii** 0.718 (3 s.f.)
b Normal approximation as n large, p close to $\frac{1}{2}$ and $np = 10 > 5$. (Exact binomial is 0.7148.)
The Poisson approximation shouldn't be used because p isn't small, it is bigger than 0.1.

11 **a** **i** A hypothesis test is where the value of a population parameter (whose assumed value is given in H_0) is tested against what value it takes if H_0 is rejected (this could be an increase, a decrease *or* a change).
ii A *range of values of a test statistic* that would lead to the *rejection of the null hypothesis*.
b Critical region $X \leqslant 3$ or $X \geqslant 16$
c 0.0432 or 4.32%
d *Insufficient* evidence to reject H_0.
The rate of incoming calls is *less* during the school holidays is *not* supported.

12 **a**

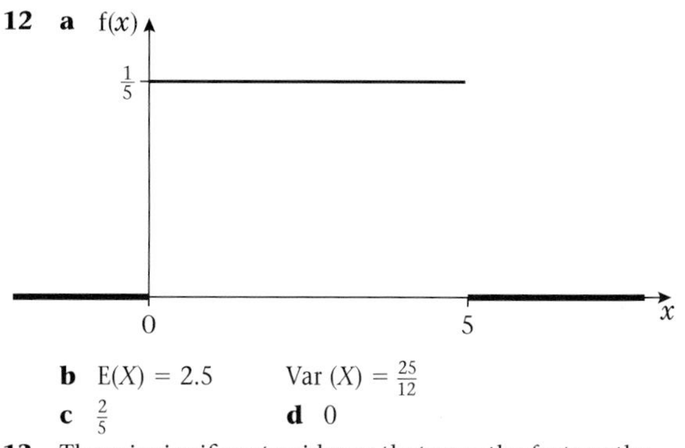

b $E(X) = 2.5$ $\text{Var}(X) = \frac{25}{12}$
c $\frac{2}{5}$ **d** 0

13 There is significant evidence that near the factory the river is polluted with bacteria at the 5% level.

14 Median 5 $= \frac{10}{40}$ or $\frac{5}{32}$
Median 10 $= \frac{54}{64}$ or $\frac{27}{32}$

15 Insufficient evidence to reject H_0.
No evidence of an increase in the number of times the taxi driver is late.

16 **a** **i** If $X \sim B(n, p)$
n is large
p is close to 0.5
and $np > 5$
then X can be approximated by $N(np, np(1-p)$
ii mean $= np$
variance $= npq = np(1-p)$
b 0.996 (3 s.f.) **c** £19 940

17 **a** A random variable that is a function of known observations from a population.
or A statistic is a numerical property of a sample.
b **i** Yes, it is a statistic.
ii No, it is not a statistic.

18 **a** 0.2225 **b** 0.2607 **c** 0.2567 **d** 0.1977

19 **a** $\lambda > 10$ or large
b The Poisson distribution is *discrete* and the normal distribution is *continuous*.
c 0.1247
d 0.1357
e 2 (or 3) Saturdays.

20 **a** $$f(x) = \begin{cases} \frac{1}{\beta - \alpha} & \alpha < x < \beta \\ 0 & \text{otherwise} \end{cases}$$
b $\alpha = -2$ $\beta = 6$ **c** 75 cm
d 43.3 (3 s.f.) **e** $\frac{60}{150} = \frac{2}{5}$

21 **a** Sufficient evidence to reject H_0.
Conclude that the proportion of family size sold is lower than usual.
b Critical region $Y = 0$ or $Y \geqslant 9$
c 3.97%

Examination Practice Paper

1 **a** **i** A named or numbered list of all members of the population.

ii A random variable consisting of **any function of the observations** and **no other quantities**. Or a numerical property of a sample.

b **i** A statistic.
Contains only observations.

ii Not a statistic.
Contains μ.

2 **a**

$$f(x) = \begin{cases} \frac{1}{4} & 1 \leqslant x \leqslant 5 \\ 0 & \text{otherwise} \end{cases}$$

b 3 **c** $1\frac{1}{3}$ **d** $\frac{3}{4}$ **e** $\frac{1}{2}$

3 **a** 0.3559 **b** 0.1694 **c** 0.159

4 **a** Any 2 from events occur
- independently of each other and at random
- singly in a continuous space or time
- at a constant rate.

b 5 to nearest day.

c 8 to nearest day.

d 22

5 **a** Critical region is $X = 0$ or $X \geqslant 7$

b 0.0607

c There is no evidence to reject the null hypothesis. The probability that a pin chosen at random is not less than 0.15.

6 **a** **i** 0.4529 **ii** 0.464

b Normal

7 **a**

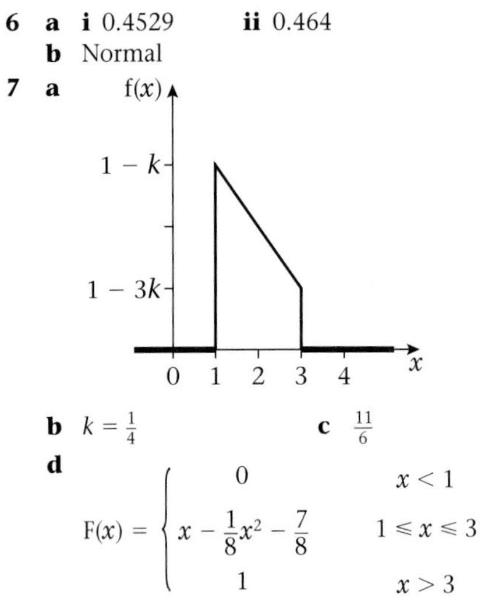

b $k = \frac{1}{4}$ **c** $\frac{11}{6}$

d

$$F(x) = \begin{cases} 0 & x < 1 \\ x - \frac{1}{8}x^2 - \frac{7}{8} & 1 \leqslant x \leqslant 3 \\ 1 & x > 3 \end{cases}$$

e 1.764

f It has a positive skew since the mean is the largest 'average'.

Index